Handbook of Fiber Science and Technology: Volume III

HIGH TECHNOLOGY FIBERS

Part B

INTERNATIONAL FIBER SCIENCE AND TECHNOLOGY SERIES

Handbook of Fiber Science and Technology: Volume III

HIGH TECHNOLOGY FIBERS

Part B

edited by

Menachem Lewin   Jack Preston

Israel Fiber Institute   *Research Triangle Institute*
and Hebrew University   *Research Triangle Park,*
Jerusalem, Israel   *North Carolina*

CRC Press
Taylor & Francis Group
Boca Raton London New York

CRC Press is an imprint of the
Taylor & Francis Group, an **informa** business

ABOUT THE SERIES

When human life began on this earth, *food* and *shelter* were the two
most important necessities. Immediately thereafter, however, came
clothing. The first materials used for it were fur, hide, skin, and
leaves—all of them sheetlike, two-dimensional structures not too abun-
dantly available and somewhat awkward to handle. It was then–quite
a few thousand years ago—that a very important invention was made:
to *manufacture* two-dimensional systems—fabrics—from simple mono
dimensional elements—fibers; it was the birth of textile industry based
on fiber science and technology. Fibers were readily available every-
where; they came from animals (wool, hair, and silk) or from plants
(cotton, flax, hemp, and reeds). Even though their chemical composition and mechanical properties were very different, yarns were made
of the fibers by spinning and fabrics were produced from the yarns by
weaving and knitting. An elaborate, widespread, and highly sophisticated art developed in the course of many centuries at locations all over
the globe virtually independent from each other. The fibers had to be
gained from their natural sources, purified and extracted, drawn out
into yarns of uniform diameter and texture, and converted into textile
goods of many kinds. It was all done by hand using rather simple and
self-made equipment and it was all based on empirical craftsmanship
using only the most necessary quantitative measurements. It was also
performed with no knowledge of the chemical composition, let alone the
molecular structure of the individual fibers. Yet by ingenuity, taste,
and patience, myriads of products of breathtaking beauty, remarkable
utility, and surprising durability were obtained in many cases. *This
first era* started at the very beginning of civilization and extended into
the twentieth century when steam-driven machinery invaded the mechanical operations and some empirical procedures—mercerization of cotton,
moth-proofing of wool, and loading of silk--started to introduce some
chemistry into the processing.

The second phase in the utilization of materials for the preparation and production of fibers and textiles was ushered in by an accidental discovery which Christian Friedrich Schoenbein, chemistry professor at the University of Basel in Switzerland, made in 1846. He observed that cotton may be converted into a soluble and plastic substance by the action of a mixture of nitric and sulfuric acid; this substance or its solution was extruded into fine filaments by Hilaire de Chardonnet in 1884.

Organic chemistry, which was a highly developed scientific discipline by that time, gave the correct interpretation of this phenomenon: the action of the acids on cellulose—a natural fiber former—converted it into a *derivative*, in this case into a cellulose nitrate, which was soluble and, therefore, spinnable. The intriguing possibility of manipulating natural products (cellulose, proteins, chitin, and others) by chemical action and thereby rendering them soluble, resulted in additional efforts which led to the discovery and preparation of several cellulose esters, notably the cellulose xanthate and cellulose acetate. Early in the twentieth century each compound became the basis of a large industry: viscose rayon and acetate rayon. In each case special processes had to be designed for the conversion of these two compounds into a fiber, but once this was done, the entire mechanical technology of yarn and fabric production which had been developed for the natural fibers was available for the use of the new ones. In this manner new textile goods of remarkable quality were produced, ranging from very shear and beautiful dresses to tough and durable tire cords and transport belts. Fundamentally these materials were not truly "synthetic" because a known natural fiber former—cellulose or protein—was used as a base; the new products were "artificial" or "man-made." In the 1920s, when viscose and acetate rayon became important commercial items polymer science had started to emerge from its infancy and now provided the chance to make *new fiber formers* directly by the polymerization of the respective monomers. Fibers made out of these polymers would therefore be "truly synthetic" and represent additional, extremely numerous ways to arrive at new textile goods. Now started the *third era* of fiber science and technology. First the basic characteristics of a good synthetic fiber former had to be established. They were: ready spinnability from melt or solution; resistence against standard organic solvents, acids, and bases; high softening range (preferably above 220°C); and the capacity to be drawn into molecularly oriented fine filaments of high strength and great resilience. There exist literally many hundreds of polymers or copolymers which, to a certain extent, fulfill the above requirements. The first commercially successful class were the *polyamides*, simultaneously developed in the United States by W. H. Carothers of duPont and by Paul Schlack of I. G. Farben in Germany. The *nylons*, as they are called commercially, are still a very important class of textile fibers covering a remarkably wide range of properties and uses. They were soon (in the 1940s) followed by the *polyesters*, *polyacrylics*, and *polyvinyls*, and somewhat later (in the 1950s) there were added the *polyolefins* and *polyurethanes*. Naturally, the existence of so many fiber formers of different chemical composition initiated successful research on the molecular and supermolecular structure of these systems and on the dependence of the ultimate technical properties on such structures.

As time went on (in the 1960s), a large body of sound knowledge on structure-property relationships was accumulated. It permitted embarkation on the reverse approach: "tell me what properties you want and I shall *tailor-make* you the fiber former." Many different techniques exist for the "tailor-making": graft and block copolymers, surface treatments, polyblends, two-component fiber spinning, and cross-section modification. The systematic use of this "macromolecular engineering" has led to a very large number of *specialty fibers* in each of the main classes; in some cases they have properties which none of the prior materials—natural and "man-made"—had, such as high elasticity, heat setting, and moisture repellency. An important result was that the new fibers were not content to fit into the existing textile machinery, but they suggested and introduced substantial modifications and innovations such as modern high-speed spinning, weaving and knitting, and several new technologies of texturing and crimping fibers and yarns.

This third phase of fiber science and engineering is presently far from being complete, but already a *fourth era* has begun to make its appearance, namely in fibers for uses *outside* the domain of the classical textile industry. Such new applications involve fibers for the reinforcement of thermoplastics and duroplastics to be used in the construction of spacecrafts, airplanes, buses, trucks, cars, boats, and buildings; optical fibers for light telephony; and fibrous materials for a large array of applications in medicine and hygiene. This phase is still in its infancy but offers many opportunities to create entirely new polymer systems adapted by their structure to the novel applications outside the textile fields.

This series on fiber science and technology intends to present, review, and summarize the present state in this vast area of human activities and give a balanced picture of it. The emphasis will have to be properly distributed on synthesis, characterization, structure, properties, and applications.

It is hoped that this series will serve the scientific and technical community by presenting a new source of organized information, by focusing attention to the various aspects of the fascinating field of fiber science and technology, and by facilitating interaction and mutual fertilization between this field and other disciplines, thus paving the way to new creative developments.

Herman F. Mark

INTRODUCTION TO THE HANDBOOK

The Handbook of Fiber Science and Technology is composed of five volumes: chemical processing of fibers and fabrics; fiber chemistry; high technology fibers; physics and mechanics of fibers and fiber assemblies; and fiber structure. It summarizes distinct parts of the body of knowledge in a vast field of human endeavor, and brings a coherent picture of developments, particularly in the last three decades.

It is mainly during these three decades that the development of polymer science took place and opened the way to the understanding of the fiber structure, which in turn enabled the creation of a variety of fibers from natural and artificial polymeric molecules. During this period far-reaching changes in chemical processing of fabrics and fibers were developed and new processes for fabric preparation as well as for functional finishing were invented, designed, and introduced. Light was thrown on the complex nature of fiber assemblies and their dependence on the original properties of the individual fibers. The better understanding of the behavior of these assemblies enabled spectacular developments in the field of nonwovens and felts. Lately, a new array of sophisticated specialty fibers, sometimes tailor-made to specific end-uses, has emerged and is ever-expanding into the area of high technology.

The handbook is necessarily limited to the above areas. It will not deal with conventional textile processing, such as spinning, weaving, knitting, and production of nonwovens. These fields of technology are vast, diversified, and highly innovative and deserve a specialized treatment. The same applies to dyeing, which will be treated in separate volumes. The handbook is designed to create an understanding of the fundamentals, principles, mechanisms, and processes involved in the field of fiber science and technology; its objective is not to provide all detailed procedures on the formation, processing, and modification of the various fibers and fabrics.

Menachem Lewin

PREFACE

This volume, like its companion volume, Part A, concentrates on fibers recently developed or potentially available commercially in the near future, rather than on high-volume commodity fibers. The emphasis in the present work, as in Part A, deals with fibers for wholly new applications. Thus, in this volume, one can find information brought together by experts in their fields on recent developments in high-modulus fibers from organic polymers or from inorganic materials, such as metals or metal oxides, and fibers for use in biomedical applications and for carrying out such functions as removal of impurities from cigarette smoke, water, and blood. The emphasis in each chapter is on the preparation, properties, and end uses of the various fibers discussed.

Because the chapters have been written by acknowledged experts in their respective fields, the truly significant developments are highlighted for the reader wishing to gain insights on the present status and the future course of development for these fibers. A future volume will deal with such subjects as optical fibers, ceramic fibers, metal fibers, and hollow fibers for separation of gases and liquids.

The editors wish to thank the contributors for sharing their expertise and for making possible this book which, it is hoped, will stimulate increased end-use applications of fibers from this rapidly expanding area of technology.

Menachem Lewin
Jack Preston

CONTRIBUTORS

Gordon Calundann Advanced Materials Department, Hoechst Celanese Corporation, R. L. Mitchell Technical Center, Summit, New Jersey

James A. Fitzgerald Textile Fibers Department, E. I. du Pont de Nemours & Company, Inc., Wilmington, Delaware

Yoshito Ikada Research Center for Medical Polymers and Biomaterials, Kyoto University, Kyoto, Japan

Michael Jaffe Research Division, Hoechst Celanese Corporation, R. L. Mitchell Technical Center, Summit, New Jersey

Yoshikazu Kikuchi* Fibers & Textiles Research Laboratories, Toray Industries, Inc., Otsu-shi, Shiga, Japan

Warren F. Knoff Fibers Department, E. I. du Pont de Nemours & Co., Inc., Richmond, Virginia

William R. Krigbaum Department of Chemistry, Duke University, Durham, North Carolina

Kichiro Matsuda Products Development Research Laboratories, Teijin Limited, Iwakuni, Yamaguchi, Japan

Shuji Ozawa† Central Research Laboratories, Teijin Limited, Hino, Tokyo, Japan

Paul G. Riewald Fibers Department, E. I. du Pont de Nemours & Company, Inc., Wilmington, Delaware

---

Present affiliations:
*Fuji Xerox Co., Ltd., Minamiashigara-shi, Kanagawa, Japan
†Prudential-Bache Securities, Minato-ku, Tokyo, Japan

James C. Romine Fibers Department, E. I. du Pont de Nemours & Company, Inc., Wilmington, Delaware

Masaharu Shimamura Fibers & Textiles Research Laboratories, Toray Industries, Inc., Otsu-shi, Shiga, Japan

Hisataka Shoji* Fibers & Textiles Research Laboratories, Toray Industries, Inc., Otsu-shi, Shiga, Japan

Michihiko Tanaka New Fibers Laboratory, Fibers & Textiles Laboratories, Toray Industries, Inc., Otsu-shi, Shiga, Japan

David Tanner Fibers Department, E. I. du Pont de Nemours & Company, Inc., Wilmington, Delaware

Kazuo Teramoto† Fibers & Textiles Research Laboratories, Toray Industries, Inc., Otsu-shi, Shiga, Japan

Hyun-Nam Yoon Research Division, Hoechst Celanese Corporation, R. L. Mitchell Technical Center, Summit, New Jersey

Toshio Yoshioka Fibers & Textiles Research Laboratories, Toray Industries, Inc., Otsu-shi, Shiga, Japan

Robert J. Young Polymer Science and Technology Group, Manchester Materials Science Centre, Manchester, England

---

Present affiliations:
*Toray Industries, Inc., Kamakura-shi, Kanagawa, Japan
†Biomaterial Research Institute Co. Ltd., Yokohama, Kanagawa, Japan

CONTENTS

CONTENTS OF OTHER HANDBOOKS

Handbook of Fiber Science and Technology: Volume III

HIGH TECHNOLOGY FIBERS

Part B

1

ARAMID COPOLYMER FIBERS

SHUJI OZAWA* / Teijin Limited, Hino, Tokyo, Japan

KICHIRO MATSUDA / Teijin Limited, Iwakuni, Yamaguchi, Japan

*Present affiliation: Prudential-Bache Securities, Minato-ku, Tokyo, Japan

1. INTRODUCTION

Fiber scientists in pursuit of an ideal synthetic fiber like to try
out new crystalline polymers with a regular sequence of a recurring
molecular unit. This is because they like to have a crystal structure
with a melting point well above the service temperature of the fiber,
and yet they have to fabricate the polymer into a fiber by means
of a solution or melt process.

    Of the three major synthetic fibers, polyester and nylon fibers
are manufactured by the melt process at ca. 250 °C, which is well
above temperatures to which fibers are often exposed and low enough
to fabricate the polymers without too much difficulty. Poly(ethylene
terephthalate), poly(hexamethylene adipamide) and poly(ε-caprolactam)
are representative polymers for the fibers, and are usually homopolymers except for special types such as those with enhanced
dyeability, high moisture uptake, etc. In the case of acrylic fibers,
however, copolymers are more common, since poly(acrylonitrile)
is not meltable nor soluble enough in a common solvent for making
good fibers. Usually, introduction of a nonrecurring monomer unit
imparts some new properties to the fiber, but often with a certain
loss in important basic properties such as the thermal and/or mechanical properties.

    There are groups of polymers in which use of copolymer is
rather common in the fiber making. Aromatic polyimides and polyarylates are the typical examples. They are mostly infusible and
insoluble when in homopolymer, and need to incorporate a third
constituent to render the polymer tractable. Since the thermal properties of the homopolymers are usually quite high, those of copolymers
are still high enough for most of the applications in which the high
temperature fibers are expected to be utilized.

    Generally speaking, the thermo-mechanical properties of copoly-
mer materials can be estimated to be somewhere between those of
the two constituent homopolymers. When the properties of a copolymer deviate significantly above or below the proportionally calculated
values, the gap is referred as the copolymer effect. Typical patterns
are shown in Fig. 1: The crystallinity and the softening point of
copolymer materials are lower, and the solubility is higher than
the proportional average of the corresponding values of the two
constituent homopolymers. This may be understood in terms of dis-
order in the molecular structure caused by the presence of the
comonomer unit.

    We can look at the properties of fabricated materials of copolymers
in the same fashion, however, with special considerations from a
different viewpoint. Here we deal with two different concepts in
a single term: for example, under tenacity of a fiber we include

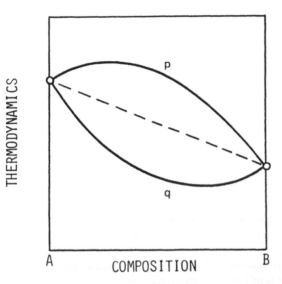

Figure 1.1 Thermodynamics versus copolymer composition. (p: e.g., solubility, tractability, q: e.g., crystallinity, softening point.)

the tenacity of real fibers as well as those calculated for a hypothetical fiber in which both the molecular structure and the molecular alignment are idealized. We would refer the latter to the theoretical value. Figure 2 is an exemplary presentation of fiber tenacity versus copolymer composition, in which S is the tenacity observable with sample fibers and T is the tenacity calculated from the bond dissociation energy at the weakest linkage along a single polymer molecule in its most extended form. It is reasonable that the S's of each homopolymer fall below the corresponding T's, but this does not necessarily mean that S's are relative to T's. In the case of fabricated material, the spatial arrangement of molecular chains is usually very much different from the idealized one, and is not determined simply by the chemical composition of the polymer chains. It is given, rather, as a result of a compound effect of the properties of the substance itself and the conditions of the fabrication process.

Now let us focus on the aramid fibers of the high modulus and high tenacity class in a modern sense. Kevlar (developmental names: Fiber B, PRD 49) was introduced as the first industrial product of this class by E. I. du Pont, and recently Twaron (developmental name: Arenka) by AKZO, and Technora (developmental name: HM-50) by Teijin. All three fibers are based largely on the para-oriented phenylene unit in their molecular structure, typically the p-phenylene

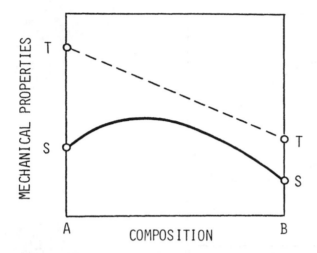

Figure 1.2 Mechanical properties of materials versus copolymer composition. ----: theoretical, ———: sample.

terephthalamide linkage. As an example, the S's and the T's are compared for poly(p-phenylene terephthalamide) [1-3]:

	Tenacity (g/d)	Tensile modulus (g/d)
S (observed)	22 [1]	1050 [1]
T (calculated)	230-300 [2]	1500 [3]

As is well known, Kevlar is the result of du Pont's success in increasing the level of the S in Fig. 2 by adoption of the liquid crystalline state in its spinning process [4-6]. Yet the S remains as a fraction of the T, suggesting that the molecular arrangement in the fiber is still far from what the scientist wished to achieve. Since the ratio S/T is quite low in tenacity at present, one cannot rule out the possibility of realizing a higher S value at an intermediate copolymer composition than those S values so far attained with each homopolymer, if better molecular arrangement is eventually provided by the copolymer effect.

2. SEARCH FOR ARAMID COPOLYMER

2.1 Fibers of Poly(p-phenylene terephthalamide) (PPTA)

The characteristic points in the manufacture of PPTA fibers were well discussed in Part A of this series [7]. The PPTA molecule is much less flexible compared to the polymers used in the major synthetic fibers; it does not melt up to 500°C, nor does it dissolve in various organic solvents. It is soluble, however, in a strong protonic solvent. It gives an optically anisotropic solution in a concentration above 8-9 wt% in concentrated sulfuric acid, and provides a spin dope of 20 wt%, suitable for spinning at 90°C, by formation of a nematic phase. By means of the dry jet-wet spinning process, even an as-spun filament usually has tensile properties as high as the high tenacity, high modulus fibers, giving a cost advantage by elimination of a familiar draw process. Some drawbacks of this fiber were notably low compression strength [8], and fatigue and wear resistance, as well as a higher tendency to fibrillate [9], all seeming to be associated with the microfibrillar structure [10-13] in the fiber. It cannot be drawn, indicating that the fiber structure, once formed, is difficult to modify by aftertreatment.

In recent years some attempts have been made to alter the crystal structure and the higher order structure in the as-spun fiber through detailed structural analyses [14,15], but essential improvement is yet to be seen.

Work was reported on drawable fibers spun from a dilute isotropic solution of PPTA. The drawing increased the axial orientation of the crystalline part to the same degree as that of the fibers from the anisotropic solution [16], but did not much increase the

Table 1.1 Polymer Design for Copolyamide Giving a High Modulus, High Tenacity Fiber

Objective	Molecular design
Thermal stability	* Wholly aromatic polyamide * Absence of unstable linkage (urethane, urea, alkylene, etc.)
Solubility	* Copolymer with dissymmetrical units * Inclusion of $-O-$, $-CO-$, $-SO_2-$, etc. * Amides rather than esters
Heat drawability	* High molecular weight * Enhanced chain flexibility by incorporating $-O-$, $-CO-$, $-SO_2-$, etc., into polymer chain
Dimensional stability	* Rigid molecular chain * Crystallinity by regular sequence in polymer

Table 1.2   Fibers from Copolymers of meta/para Configuration

Composition (mol. ratio)				I.V. of polymer[a]	Spinning solution[b] (wt.%/solvent)
$-N\langle\bigcirc\rangle-N-$ (H,H)	$-N\langle\bigcirc\rangle-N-$ (H,H)	$-C\langle\bigcirc\rangle-C-$ (O,O)	$-C\langle\bigcirc\rangle-C-$ (O,O)		
100	0	100	0	5.5	$20/H_2SO_4$ (90°C, aniso.) $10/H_2SO_4$ (RT, aniso.)
85	15	100	0	4.6	$6.0/\dfrac{HMPA}{NMP} + LiCl$
80	20	100	0	4.4	$9.0/\dfrac{HMPA}{NMP} + LiCl$
50	50	100	0	1.9	$8.5/\dfrac{HMPA}{NMP} + LiCl$
0	100	100	0	1.3	$14.0/\dfrac{HMPA}{NMP} + LiCl$
75	25	100	0	1.43	$10.0/HMPA + LiCl$
100	0	75	25	(missing)	$10.0/HMPA + LiCl$
50	50	50	50	1.46	$10.0/\dfrac{HMPA}{NMP} + LiCl$

[a]I.V. = inherent viscosity, at 30°C with 0.5 g polymer/100 ml H_2SO_4.
[b]HMPA = hexamethylphosphortriamide; NMP = N-methyl-2-pyrrolidone.
[c]D = denier per filament; T = tensile strength; E = elongation to break; M = tensile modulus.
[d]DR = draw ratio.

tenacity and the elongation at break, presumably due to macroscopic defects such as porosity in the fiber structure.

With all of the above knowledge in mind, the present authors arrived at the ideas summarized in Table 1 [17] as to ways to bring the S values of copolymer fibers closer to the T values. In our copolymer study, new diamines were chosen rather than new diacid components because of preparative ease. Our prime interest was to identify a workable copolymer system that would be soluble in an organic solvent. The following discussion is based on those results.

Spinning method (coagulant)	Hot drawing[d] (°C/DR)	Fiber properties[c]				Ref.
		D (d)	T (g/d)	E (%)	M (g/d)	
dry jet-wet (H$_2$O)	none (as spun)	1.5	25	4	500	1,6
	500/1.05	1.5	22	3	1000	1,6
wet (H$_2$O)	none (as spun)	4.0	7.4	5.6	298	18
	500/1.03	3.9	14.0	1.9	940	18
wet (RT, 40% CaCl$_2$ aq.)	400/1.07	7.1	6.0	3.4	365	18
wet (RT, 40% CaCl$_2$ aq.)	450/1.45	1.3	4.7	2.5	260	18
wet (90°C, 45% CaCl$_2$ aq.)	330/2.10	2.2	8.4	6.2	260	18
wet (90°C, 45% CaCl$_2$ aq.)	325/1.90	5.1	6.6	11.6	101	18
wet (RT, H$_2$O)	500/1.61	3.1	5.9	1.6	425	19
wet (RT, H$_2$O)	300/1.4 + 500/1.0	2.0	6.1	1.5	450	19
wet (RT, H$_2$O)	300/1.7 + 500/1.0	1.8	5.3	2.1	326	19

2.2 Introduction of the Meta Configuration [18]

The results of the copolymer study are summarized in Table 2, which also includes a similar study by Unitika [19]. The point is to see the effect on the fiber properties by partially replacing the para-phenylene group with the meta analog in the otherwise straight backbone of the PPTA chain. The conclusive remarks on these works are that the tensile properties of the whole copolymer fibers ranged between those of the fibers of PPTA and poly(m-phenylene isophthalamide) (PMIA), and there was no unusual copolymer composition at which unexpected fiber properties were observed.

2.3 Copolymers of Isomorphism

2.3.1 Methyl Substitution on the Phenylene Ring

A series of studies was made to see the effect of a small substituent on the terephthaloyl moiety in PPTA on the properties of the fiber thereof. However, expected improvement in the solubility of the copolymers was not significant, even after 30% replacement with a methylterephthaloyl group in PPTA [20]. As shown in Table 3, the fiber properties are not very attractive either.

2.3.2 Methyl Substitution at the Amide Linkage

In the hope of improving solubility by reducing the intermolecular hydrogen bonding, studies were made on copolymerization with N-methyl p-phenylenediamine [21]. The copolymers became soluble in an organic solvent, and their sulfuric solution did not show optical anisotropy even at high concentration. The as-spun fibers from an organic solution showed some drawability, but the tensile properties of the resultant fibers, including poly(N-methyl p-phenylene terephthalamide), were rather poor, although the as-spun fiber of the latter exhibited significant drawability as shown in Table 4.

2.4 Copolymers with Large Diamines

Incorporation of diamines with wider distances between the two amino groups would reduce both the spatial density of the hydrogen bonding and the rate of crystallization. This would be expected to improve the processability of the copolymers. The results of studies along this line are summarized in Table 5. The present authors did not expect too much from the results, because the copolymerization of this kind would adversely affect the crystallinity of the polymer chain before the solubility is practically improved. As shown in the table, those as-spun fibers of copolymers with 3,4'-diaminodiphenyl ether(3,4'-ODA) seemed to be highly drawable over a wide range of copolymer composition; surprisingly, the fibers spun from a homogeneous isotropic solution in an organic solvent and highly drawn at a high temperature did show tensile properties comparable to the PPTA fibers from the liquid crystalline dope. Also interesting was that these unusual copolymer effects seemed observable only in the case of 3,4'-monomers and not in the case of 4,4'-monomers [22,23].

   The authors' experience with aramid copolymers can be summarized as below:

Table 1.3 Fibers from Methyl Substituted Copolymers on the Phenylene Ring[a]

Composition (mol. ratio)		I.V. of Polymer	Spinning solution (wt.%/solvent)	Spinning method (coagulant)	Hot drawing (°C/DR)	Fiber properties			
H–N–⟨⟩–N–	O=C–⟨CH₃⟩–C=O / O=C–⟨⟩–C=O					D (d)	T (g/d)	E (%)	M (g/d)
100	15	5.98	12/H_2SO_4 (aniso.)	wet (RT, H_2O)	500/1.03	3.7	14.7	2.1	810
100	30	5.43	13/H_2SO_4 (aniso.)	wet (RT, H_2O)	500/1.03	4.2	14.7	1.9	897
100	100	3.30	16/H_2SO_4 (aniso.)	wet (RT, H_2O)	480/1.0	6.8	8.3	1.9	727
100	100	3.30	11/$\frac{HMPA}{NMP}$ + LiCl (iso.)	wet (50°C, $CaCl_2$ aq.)	480/1.0	5.1	3.6	0.9	561
100	0	5.49	10/H_2SO_4 (aniso.)	wet (RT, H_2O)	500/1.03	2.7	14.0	1.9	940

[a]See Table 2 for abbreviations.

Table 1.4 Fibers from Methyl Substituted Copolymers at the Amide Linkage[a]

Composition (mol. ratio)		I.V. of Polymer	Spinning solution (wt.%/solvent)	Spinning method (coagulant)	Hot drawing (°C/DR)	Fiber properties			
H H / –N–⟨⟩–N–	CH_3 / –N–⟨⟩–N–	O=C–⟨⟩–C=O				D (d)	T (g/d)	E (%)	M (g/d)
85	15	100	2.8	10/$\frac{HMPA}{NMP}$ + LiCl (iso.)	wet (RT, 30% $\frac{HMPA}{NMP}$ aq.)	500/1.27	2.2	5.5	2.1	354
50	50	100	1.9	14/$\frac{HMPA}{NMP}$ + LiCl (iso.)	wet (RT, 43% $CaCl_2$ aq.)	350/1.38	5.9	3.6	2.1	220
0	100	100	2.1	16/$\frac{HMPA}{NMP}$ + LiCl (iso.)	wet (RT, 40% $CaCl_2$ aq.)	boiling water/3.85 + 360/1.53	2.0	10.0	4.8	442

[a]See Table 2 for abbreviations.

Table 1.5 Fibers from Terephthalic Copolymers with Large Diamines

Diamine Composition (mol.%)			I.V. of Polymer	Spinning solution[a] (wt.%/solvent)
H H $-N-\bigcirc-N-$	H H $-N-X-N-$			
85	X: $-\bigcirc-CH_2-\bigcirc-$,	15	2.8	$10/H_2SO_4$
85	$-\bigcirc-SO_2-\bigcirc-$,	15	4.8	$7/\dfrac{HMPA}{NMP}$ + LiCl
70	$-\bigcirc-SO_2-\bigcirc$,	30	3.1	6.8/HMPA
50	$-\bigcirc-SO_2-\bigcirc$,	50	2.1	10.5/DMA
0	$-\bigcirc-SO_2-\bigcirc$,	100	1.8	15.8/NMP
85	$-\bigcirc-O-\bigcirc-$,	15	4.3	$6/\dfrac{HMPA}{NMP}$ + LiCl
70	$-\bigcirc-O-\bigcirc-$,	30	4.1	$6/\dfrac{HMPA}{NMP}$ + LiCl
0	$-\bigcirc-O-\bigcirc-$,	100	3.6	$7/\dfrac{HMPA}{NMP}$ + LiCl
70	$-\bigcirc-O-\bigcirc$,	30	2.5	6.4/NMP
50	$-\bigcirc-O-\bigcirc$,	50	2.8	6.0/NMP
0	$-\bigcirc-O-\bigcirc$,	100	2.4	$9.6/\dfrac{HMPA}{NMP}$ + LiCl
70	$-\bigcirc-O-\bigcirc-O-\bigcirc-$,	30	2.8	$5/\dfrac{HMPA}{NMP}$ + LiCl
0	$-\bigcirc-O-\bigcirc-O-\bigcirc-$,	100	2.3	$12/\dfrac{HMPA}{NMP}$ + LiCl

[a]DMA = dimethylacetamide. See Table 2 for other abbreviations.

1. Any copolymers whose copolymer ratio is below 15 mol% do not seem to give polymers that are basically different from the homopolymer, although some changes in the properties of the polymer solution, spinnability, and fiber properties may be observed.
2. Incorporation of comonomers in which two phenylene groups are connected with a single atom unit generally improves the fibers' tensile properties by drawing.
3. Maximum effects seem to be observed with the 3,4'-diamines and perhaps with 3,4'-diacyloyl compounds.

Spinning method (coagulant)	Hot drawing (°C/DR)	Fiber properties			
		D (d)	T (g/d)	E (%)	M (g/d)
wet (RT, H_2SO_4 aq.)	400/1.05	4.5	4.0	11.2	130
wet (RT, $CaCl_2$ aq.)	400/1.50	4.2	3.4	6.3	105
dry jet-wet (95°C, $CaCl_2$ aq.)	430/6.2	1.8	15.0	3.5	445
dry jet-wet (95°C, $CaCl_2$ aq.)	430/7.3	1.1	16.6	4.2	438
dry jet-wet (95°C, $CaCl_2$ aq.)	390/4.5	2.6	9.8	3.3	277
dry jet-wet (RT, $CaCl_2$ aq.)	450/1.5	2.0	2.0	5.2	55
dry jet-wet (RT, $CaCl_2$ aq.)	450/2.5	1.6	12.5	4.0	375
dry jet-wet (RT, $CaCl_2$ aq.)	450/1.6	1.9	12.1	10.5	95
dry jet-wet (70°C, $CaCl_2$ aq.)	510/6.0	2.1	24.5	4.6	622
dry jet-wet (70°C, $CaCl_2$ aq.)	495/10.0	1.8	27.3	5.0	625
dry jet-wet (70°C, $CaCl_2$ aq.)	390/4.25	1.4	12.7	4.6	324
dry jet-wet (RT, $CaCl_2$ aq.)	460/1.3	1.8	10.0	3.6	330
dry jet-wet (90°C, $CaCl_2$ aq.)	460/1.5	2.4	8.6	6.5	172

Examples of the aramid of dissymmetrical monomers were reported by Bayer A.G. to improve the tractability, and are summarized in Fig. 3. The works are on various homopolymers, giving fibers of high modulus but rather low tenacity. They also showed that the preferred molecular skeleton has at least one meta orientation in phenylene linkages. Work on poly(3,4'-ODA terephthalamide) was performed in the USSR [25], but no work was reported on its fibers. Some detailed work by the present authors on the copolymer of PPTA with 3,4'-ODA will be discussed in the next section.

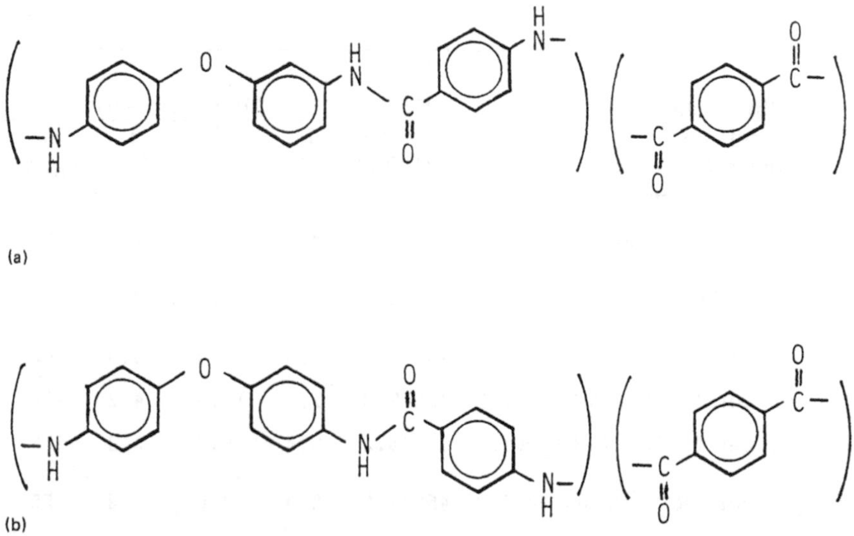

(a)

(b)

Figure 1.3 Bayer's aramid fiber. (a) Fiber, wet spun and drawn (×6.0). T = 7.2 g/dtex, E = 2.0%, M = 9200 kg/mm². (b) Fiber, wet spun and drawn (×2.0). T = 2.9 g/dtex, E = 3.5%, M = 900 kg/mm². (From Ref. 24.)

3. COPOLYTEREPHTHALAMIDES OF p-PHENYLENEDIAMINE AND 3,4'-ODA

3.1 The Effect of the Diamine Composition

Some data previously shown in Table 5 demonstrate the effects of the diamine composition on the copolyterephthalamides of p-phenylenediamine and 3,4'-ODA, on the appropriate conditions for fiber making, and on the properties of the resultant fibers. The results of more detailed laboratory experiments are presented in Fig. 4 to give a comprehensive picture of the copolymer effects [23]. This figure clearly indicates that the tensile properties of the drawn fibers are maximized in the middle, just as the authors initially anticipated.

For simplicity and convenience, the authors will hereafter assign codes for the copolymers to indicate the content of 3,4'-ODA in the diamine composition in mole percent such that P-40 is a copolyterephthalamide of 40 mol% of 3,4'-ODA and 60 mol% of p-phenylenediamine.

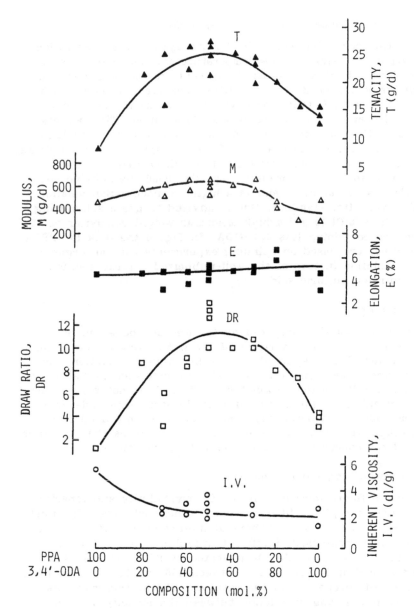

Figure 1.4 Effect of molar ratio of diamine components on polymer
I.V., DR, and fiber properties of P-series copoly(terephthalamide).

3.1.1 Polymer Concentration of the Spin Dope

The homopolymer P-100 can easily be prepared in a concentration
of 10 wt% by the standard low temperature solution polycondensation
in dry N-methyl pyrrolidone as the solvent from an equivalent amount
of terephthaloyl chloride and 3,4'-ODA. The reaction mixture after
neutralization with $Ca(OH)_2$ or CaO can serve as the spin dope
with usual pretreatment: filtration and de-aeration. With decrease
in the 3,4'-ODA content, a reduced dope concentration was found
most suitable: ca. 6% for P-40 to -50. Those spin dopes are optically
isotropic and stable, and preferably have viscosities of the order
of some tens to some hundreds poises. For P-30 the suitable concentration was 5% at best, and in a higher concentration gelation
was observed after standing. One is advised to use a solubilizing
agent such as LiCl to get a high molecular weight copolymer of
P-30 or one containing less 3,4'-ODA. In Fig. 4 the data for PPTA,
namely P-00, are based on spinning experiments with an organic
isotropic solution instead of the well-publicized nematic solution
in concentrated sulfuric acid.

3.1.2 Spinning

The copolymer solution prepared by the method above can be subjected to fiber spinning by means of the wet or dry jet-wet spinning
to give a workable as-spun fiber filament. Aqueous solutions containing the same solvent less than 50 vol% or a salt such as $CaCl_2$ were
used successfully as the coagulation medium. It is not very difficult
to find an effective spinning conditions through simple trial-and-
error methods by changing the amount and the kind of additive
to the coagulation bath, and the bath temperature as the copolymer
ratio and the concentration of the spin dope are varied.

3.1.3 Drawing and Fiber Tenacity

The as-spun fibers were usually low in crystallinity and tenacity
(ca. 2 g/d), except for fibers of the two homopolymers. The most
characteristic feature of the fibers of intermediate composition was
that the fibers possessed significant drawability at a rather high
temperature. The drawability is indicated as DR in Fig. 4, and
was determined at a temperature between 450 and 500°C. Also plotted
are the tensile properties of the resultant fibers. One may note
that there is a strong correlation between the tenacity and the
draw ratio, and that the tensile properties are comparable to Kevlar
only for the copolymer fibers from P-30 to P-70, and not for P-05
or P-95.

3.2 Discussion of the Drawing Mechanism of the Copolymer Fibers

The success of drawing the copolymer fibers may need some further discussion. Although the copolymers consist of more flexible molecular chains than PPTA (or P-00), the chain may not be flexible enough to form the folded-chain lamellae, familiar in common fibers such as nylons and polyesters. In other words, the copolymer chain is more properly said to be only semirigid or semiflexible. What is puzzling is that the copolymer fibers showed high drawability far beyond the extent known for nylons and polyesters.

The fiber properties and structural parameters for the fibers are summarized in the figures [26]: Fig. 5 is for the fibers drawn on a hot plate to 70% of the maximum draw ratio, at which the fiber ruptured at various temperatures; Fig. 6 is for the fibers drawn to various ratios in the same fashion at 495°C.

3.2.1 The Effect of Drawing Temperature

The copolymer P-50 does not show a clear melting point, and starts to decompose before it gives a clear melt. Upon heating the polymer in a cylinder of a capillary rheometer (Koka Flow Tester, made by Shimazu) under pressure exerted by a plunger, the content started to flow out through an orifice at a temperature between 450 and 470°C. We referred to this temperature as the "quasi-polymer melt temperature" and indicated it as Tm' in Fig. 5 [27].

As shown in Fig. 5, sample draw ratio (DR) of the as-spun fiber increases significantly at temperatures above Tm', as does the tenacity of the drawn fibers. The elongation at break of the drawn fibers, on the other hand, drops sharply at a much lower temperature—between 300 and 350°C—and does not change much beyond that temperature. Interestingly, the structural parameters derived from X-ray diffraction seem to behave differently: The apparent degree of crystallinity, X, the degree of crystal orientation, F, and the apparent size of the crystallite, D_{200}, seem to increase gradually, and reach a plateau, for X at ca. 450°C and for D_{200} at ca. 390°C. We will return to discuss on D_{200} in the next section.

The degree of crystal orientation, F, was derived by the following formula with use of the half peak width H (in degrees) of the azimuthal scan of the 200 plane reflection (at $2\theta = 20.3°$ with Cu-Kα) observable in crystalline PPTA fibers:

$$F(\%) = \frac{100 \times (180° - H)}{180°}$$

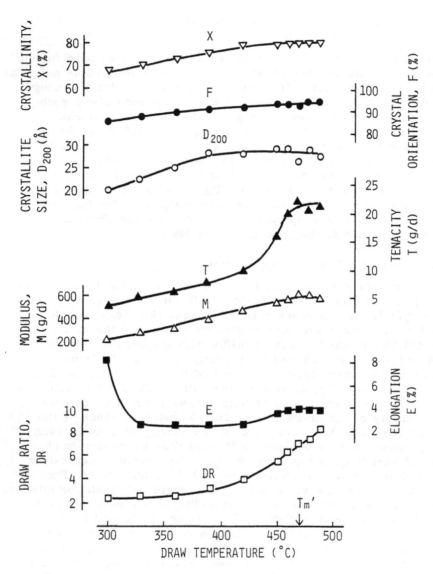

Figure 1.5 Draw ratio, tensile properties, and apparent structural parameters of P-50 fiber as functions of draw temperature. Drawn with DR = 0.7 × (maximum DR).

The apparent crystallite size D_{200} was calculated from the half peak width B_{200} of the equatorial scan of the 200 plane reflection of PPTA by Scherrer's equation:

$$D_{200} (\text{Å}) = \frac{K \lambda}{B_{200} \cdot \cos \theta}$$

where K is a correction factor of the order of unity, λ is the wavelength of the X-ray (1.54 Å), and θ is the Bragg angle.

3.2.2 The Effect of Draw Ratio

Drawing the as-spun fibers of P-50 on a hot plate at an intermediate temperature, e.g. 495°C, that is above the quasi-polymer melting temperature, Tm', and below the polymer decomposition temperature seems to reveal an interesting behavior of the fiber, as shown in Fig. 6, where the data in parentheses at DR = 1.0 are those of the as-spun fibers themselves, which are different from others receiving no thermal treatment.

The tenacity of the drawn fibers increases with DR to give the maximum value at around DR 14, and decreases by further drawing. The initial modulus of the drawn fibers varies in a similar fashion but to much less extent, although it reaches its maximum value at a lower DR. To our surprise, the elongation at break also increases continuously with DR in the same fashion as does the tenacity to its maximum at DR 15. In other words, drawing gives the resultant fibers both extensibility and stiffness at the same time.

Now let us look at the parameters for X-ray diffraction. The apparent degree of crystallinity, X, rises sharply from a low value of ca. 20% to a steady 83% at DR 6 or higher, and in Fig. 5 it gradually increases with the draw temperature. The degree of crystal orientation, F, changes in the same manner. These behaviors are rather common with fibers from ordinary crystalline polymers. A characteristic change, however, is seen in the apparent crystallite size, D_{200}. It sharply decreases from 44 Å of the as-spun fiber to a steady 30 Å by drawing above a DR of 6 at a constant temperature of 495°C. In addition, in Fig. 5, D_{200} gradually increases with the draw temperature up to ca. 30 Å at 390°C from ca. 20 Å at 300°C. However, it should be noted that the sample draw ratio of the fibers here is varied as a function of the draw temperature. This means that at 300°C, the D_{200} of 44 Å of the as-spun fiber dropped to 20 Å upon drawing only by 2.5 times.

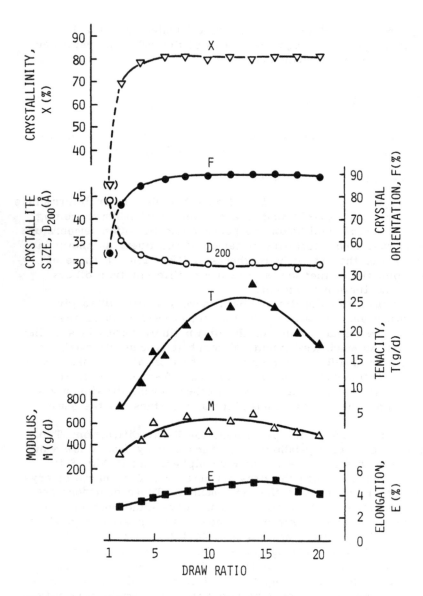

Figure 1.6   Tensile properties and apparent structural parameters of P-50 fiber as functions of draw ratio. Drawn at 495°C.

3.2.3 Discussion

There are three different mechanisms assumed to exist in the drawing process of P-50 fibers:

1. The ordinary draw mechanism seems to dominate for DR < 2
at 495°C, characterized by the increase in the orientation of polymer
molecules and in crystallization. The crystal structure weakly observable in the as-spun fiber is most likely a quasi-crystal consisting
of distorted lattice, due probably to the irregular monomer sequence
along the copolymer chain. Upon drawing, spatial molecular rearrangement re-forms the quasi-crystal into smaller and segmented but more
regular crystallites.

2. The intercrystallite shearing mechanism seems to set in
for DR = 2-14, characterized by local chain slip in the most stressed
part of the oriented and amorphous region between crystallites.
Elimination of imperfection in the structure gives rise to significant
increases in modulus and elongation, and as a consequence in the
tenacity of the fibers.

3. The excessive draw mechanism seems to become visible for
DR ≥ 15, characterized by the formation and spreading of structural
faults located in the amorphous region where molecular chain ends
or branches are concentrated even in a highly oriented molecular
assembly. Development of this fault could be observed as the formation of micro-voids or micro-fibrils. If this fault takes place, the
modulus, the elongation, and the tenacity of fibers would start
to decrease.

Mechanism 2 above may be the same as the superdraw recently
often referred to with polyolefin fibers [28]. Figure 7 is designed
to visualize the conceptual correlation between the industrial processes in fiber making and the associated changes in the molecular
process, by which the present authors wish to comprehend how
the fibers of semirigid or flexible polymers can attain the high
tenacity and the tensile modulus of the fibers of rigid polymers.
It should be noted that the rigid polymers of commercial interest
may be limited in number because of their chemistry, but the semirigid or flexible polymers offer much wider choices.

From the viewpoint discussed above, Bayer's fibers shown in
Fig. 3 [24] and Monsanto's fibers of the copolymers shown below
[29] may be categorized as following the semirigid polymer approach,
achieving the high tenacity or the high modulus by high ratio drawing:

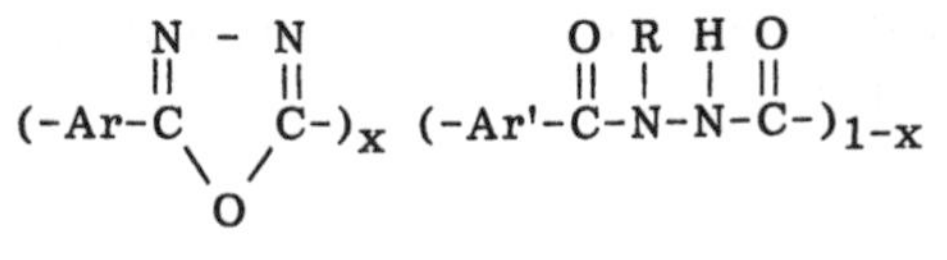

$$(-Ar-\underset{\substack{\| \\ O}}{C}\overset{\substack{N-N \\ \| \ \|}}{\diagdown\diagup}C-)_x \ (-Ar'-\overset{\substack{O \ R \ H \ O \\ \| \ | \ | \ \|}}{C-N-N-C-})_{1-x} \qquad R: \text{lower alkyl}$$

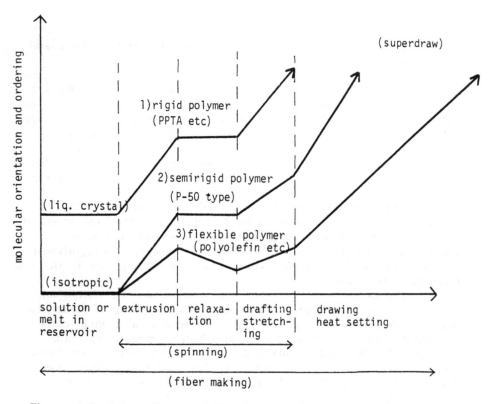

Figure 1.7 Schematic correlation between industrial process and molecular ordering process.

Now let us discuss the reasons why the fibers of copolytereph-thalamide of 3,4'-ODA were successful in achieving a high performance fiber. It is well known that the presence of the ether linkage in the polymer backbone enhances the tractability. In the case of 3,4'-ODA, a few other factors should be taken into consideration:

1. Maximum or average cross-sectional area of the molecule would be reflected in the tenacity and the modulus: 3,4'-ODA terephthalamide is much slimmer than 4,4'-ODA terephthalamide when extended along the longer axis.
2. The geometric dissymmetry of 3,4'-ODA further reduces the regularity of the backbone polymer chain even of a given copolymer composition, and consequently enhances the tracta-bility by reduction both in the degree of crystallinity and the rate of crystallization.

Therefore, the copolymer P-50 must have the best balance of chain flexibility and crystallizability, and eventually becomes capable of high ratio drawing to give fibers of high tensile properties.

Jinda and Kawai [30] have demonstrated high ratio drawing of the as-spun fibers of copolyterephthalamides of 2-chloro-p-phenylenediamine. Their data are summarized in Fig. 8 to show that the copolyamide containing the ether linkage were most drawable, and consequently gave the maximum tenacity and elongation at break.

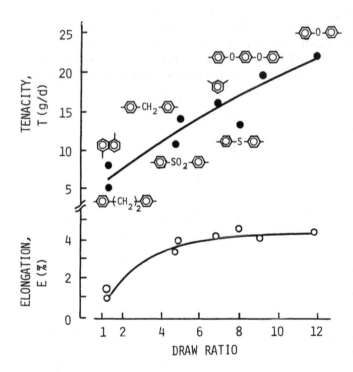

Figure 1.8 Tensile properties of the fibers made from various copoly(terephthalamide) as functions of draw ratio. Chemical symbol beside each point shows "X" in

(Data from Ref. 30.)

4. TECHNORA® FIBER

Based on knowledge of the fibers of copolyterephthalamide P-50, Teijin has decided upon commercialization of Technora, which has quite high resistance against hydrolysis in addition to the expected high tensile properties. The characteristic properties of the fiber and their associated discussions [31] are introduced in this section.

4.1 Manufacturing Process

The manufacturing process of Technora [31,32], shown schematically in Fig. 9, fully utilizes the advantage of improved tractability of P-50. Para-phenylenediamine (PPA) and 3,4'-ODA are reacted with terephthaloyl chloride (TPC) in an amide solvent such as N-methyl pyrrolidone to complete the polycondensation. The reaction mixture

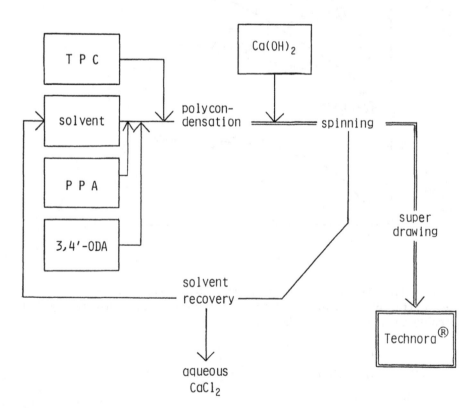

Figure 1.9   Technora process.

Table 1.6 Typical Properties of Technora Fiber

Tensile properties:		
Density	g/cm^3	1.39
Tenacity	g/d	25
	kg/mm^2	310
Tensile modulus	g/d	570
	kg/mm^2	7100
Elongation at break	%	4.4
Thermal behavior:		
Melting point	°C	>500 (decomp.)
Thermal shrinkage at 300°C	%	<0.5
Tenacity at 200°C	g/d	16
Tensile modulus at 200°C	g/d	490
Flammability	L.O.I.	25
Chemical resistance: (tenacity retention after 100 hr of exposure)		
Air (200°C)	%	95
Water (120°C)	%	97
40% H$_2$SO$_4$ aq. (95°C)	%	89
10% NaOH aq. (95°C)	%	84

is subsequently neutralized to give a stable viscous solution of P-50, which is subjected to spinning into an aqueous coagulation bath. The as-spun fiber thus formed is brought to extraction of the residual solvent, superdrawn at a high temperature, and passed through finishing to give the final product.

Since only a single solvent is used throughout and there is no isolation of the polymer, the whole process becomes very simple and straightforward. In comparison with other industrial processes for aramid fibers, no use of sulfuric acid not only greatly simplifies the solvent recovery process, but also makes Technora completely free of unfavorable residual acid problems.

4.2 Mechanical Properties

Typical tensile properties and their temperature dependence are given in Table 6 and in Fig. 10 [33]. The levels of those properties can be said to be comparable to those of highly crystalline PPTA fibers over the temperature range of practical use. Of the dynamic performance of Technora, the tenacity retention of a twine cord after 2000 times of bending under load and the number of abrasive

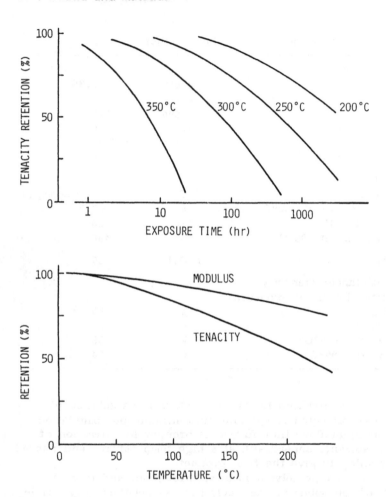

Figure 1.10 Heat (in air) resistance of Technora fiber.

cross-strokes until cord rupture are shown in comparison with a
PPTA fiber in Table 7 [34]. By microscopic examination the difference
between the two fibers seems to be connected with the tendency of
fibrillation.

The comparatively high fatigue resistance of Technora may
derive from the flexibility of the copolymer chain and presumably
rather loose crystal structure in the copolymer. Figure 11 shows
the X-ray fiber diagrams of Technora and a PPTA fiber. The PPTA
fiber gives a clearer and more detailed WAXS diffraction pattern
than does the copolymer fiber, reflecting its well-developed crystal

Table 1.7 Fatigue and Wear Resistance

Cord[a]	Tenacity retention after bending[b]	Frictional wear resistance[c]	
		Against steel	Against itself
Technora	52%	884	287
PPTA	36%	234	120

[a]Cord: twine (twist 40/40 turns/10 cm), 1500 de × 2.
[b]Tenacity was measured after 2000 times of bending under tension
(0.6 g/d), (bend radius/cord radius): 15.
[c]Number of abrasive strokes until rupture under load (600 g).

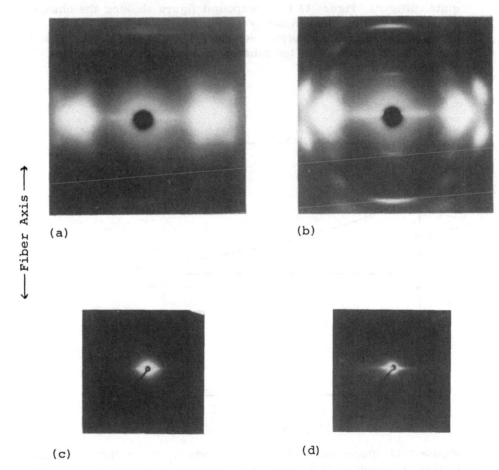

Figure 1.11 X-ray diffractions of Technora and PPTA fibers: (a) Technora, WAXS, (b) PPTA, WAXS, (c) Technora, SAXS, (d) PPTA, SAXS.

structure. Another characteristic difference may be seen in the SAXS patterns. It can be said that the PPTA fiber gives a streak more extended in the equatorial direction and that its oriented crystallites are discontinued sidewise at a certain interval, indicating the presence of microfibrillar structures.

4.3 Chemical Properties

Other characteristic features of Technora are well represented in its chemical resistance. Figure 12 shows the change in inherent viscosity of three typical aramid polymers dissolved in 98% sulfuric acid [31], indicating that P-50, incorporating the ether linkage in its polymer chain, was more susceptible to strong acids. However, the hydrolytic resistance of P-50 in the form of fibers appeared quite different. Figure 13 is a compound figure showing the change in tenacity and inherent viscosity of the two fibers immersed in an acidic or alkaline hydrolytic medium [31]. The PPTA fiber rapidly lost both its tenacity and the inherent viscosity, while Technora remains almost unchanged.

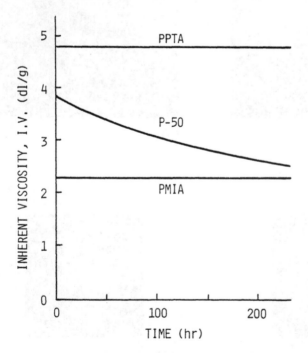

Figure 1.12   Degradation of aramids in 98% H_2SO_4 at 30°C (concentration: 0.5 g/dl). PMIA: poly(m-phenylene isophthalamide).

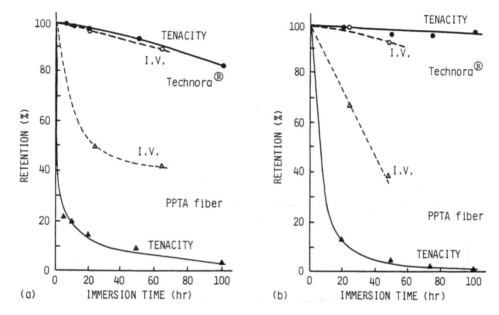

Figure 1.13 Tenacity retention and degradation of aramid fibers in hydrolyses (95°C immersion): (a) in 10% NaOH aq., and (b) in 20% H_2SO_4 aq.

The similar difference in the behavior of the two fibers was also found in resistance against neutral water under various conditions, and is summarized in Fig. 14 [35].

Figure 15 is still another demonstration of the difference of the two fibers. It shows the results of an immersion test in aqueous hypochlorite solution [31]. The PPTA fiber, only after 5 hr of immersion in a 10% aqueous solution of NaClO, lost 10% of its original weight and most surprisingly 70% of its initial strength, while only a slight change was observed in the inherent viscosity of the residual fiber. This seems to imply the presence of a heterogeneous structure that develops cylindrically along the fiber axis, consisting of a rather thin surface layer and a thick core, the former being very strong mechanically but highly susceptible to chemical attack and the latter contributing little to the mechanical properties of the fiber but rather stable chemically. For readers' interest, if one calculates the strength of the lost part of the fiber, one finds it to be of the order of 150 g/d [$T = (0.7/0.1)T_0$, where T_0 = 22 g/d].

On the other hand, the tenacity curve of Technora fiber starts to deviate from the diagonal line with immersion longer than ca. 70 hr with 20% loss both in strength and weight, suggesting rather dense and uniform texture that degrades rather slowly.

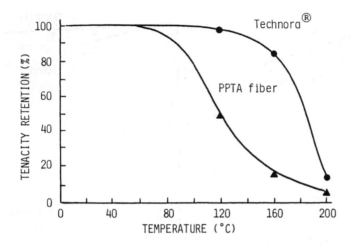

(a)

(b)

Figure 1.14 Hydrolytic resistance of yarns immersed in water: (a) 100 hr immersion, and (b) 120°C immersion.

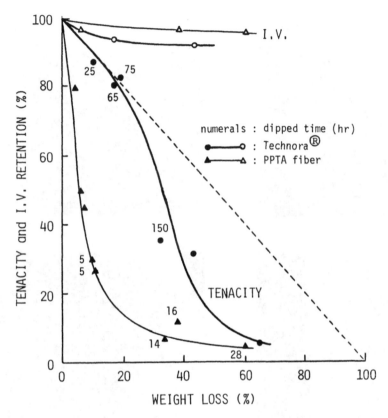

Figure 1.15 Tenacity retention and degradation of aramid fibers dipped in 10% NaClO aq., 95°C.

4.4 Structural Considerations

Morphological studies on the surface structure of PPTA fibers have been made by several groups using SEM observations after chemical etching [11,37-39]. However, they seem to have a rather common image that the supermolecular structure of the PPTA fiber consists of two distinct parts, a surface skin layer and an internal core. In the skin layer, each molecular chain is uniformly oriented and distributed; while in the core, the main body of a fiber, the structure consists of an assembly of highly oriented microfibrils, which are substructured by longitudinal sequence of crystallites, being of the order of extended PPTA molecules in length and therefore connected by limited number of the tie molecules. In other words, at each boundary the crystallite is surfaced by the chain ends

⟺ 1μm   ⟺ 1μm

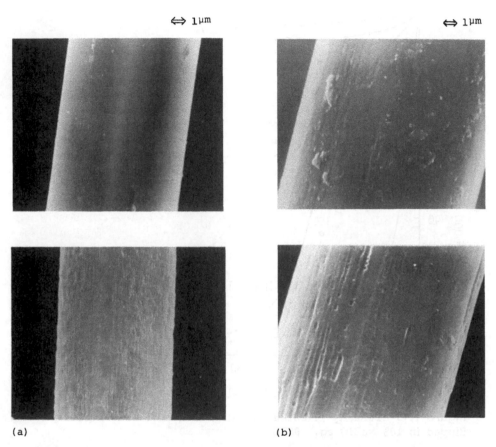

(a)   (b)

Figure 1.16 SEM images of aramid fiber surfaces. Top, blank, and bottom, chemically etched surfaces treated with 10% NaClO aq. at 95°C, 1 hr of (a) PPTA fiber, (b) P-50 fiber.

except for a few tie-molecules, which are normally under high stress. Therefore, the microfibrils are vulnerable to chemical attacks. The SEM images of the fiber surface chemically etched by a hypochlorite solution [39] are shown in Fig. 16a.

These models remain hypotheses deduced from various observations in the structural analyses, and there is no unanimous explanation about how and why such structures are resulted. It may be understood that the nematic structure in the spin dope is re-formed into the smectic structure during the course of coagulation, and this may explain the presence of a periodic sequence of structure

of the order of the average molecular length in the core of the
fiber. Morgan et al. explained the core structure as being caused
by strong interaction of the solvent acid and the chain ends covering the crystallites [36]. As those researchers pointed out, the
connecting portion between the crystallites seems to be quite weak,
both mechanically and chemically. The present authors recognized
that when processed in the same way as the PPTA fiber, a fiber
from a p-phenylene terephthalamide copolymer with 4,4'-diphenoyl
chloride much improved both the mechanical and the chemical properties, suggesting less chances of forming the smectic structure in
this copolymer system [38].

In case of Technora, on the other hand, the molecular chains
are so flexible and the crystallites so loose in structure that the
internal stresses developed upon drawing may have been mostly
dissipated during the process. Thus, there will be left only a less
distinct boundary around the loose crystallites. This understanding
is supported by X-ray diffraction (Fig. 11) and also by the SEM
image of the fiber surface etched with a hypochlorite solution
(Fig. 16b).

Moisture uptake of a fiber is caused by the hydrophilicity of
the constituent polymer, namely by the content of hydrophilic groups
and the higher order of structure, primarily the degree of crystallinity. In an atmosphere of 65%RH, Technora absorbs moisture to
2% of its weight, while a PPTA fiber spun from an anisotropic dope
absorbs to 4-6% [40]. If one assumes water molecules to be absorbed
only by the surface amide groups, the above data means that PPTA
fibers have more than twice as much wide internal surface as does
Technora. And if this internal surface is derived from the micro-
fibrils, one may further explain the difference of the two fibers
in chemical resistance as well as in wear and fatigue resistance.

Assuming that the high chemical resistance of Technora is provided only by uniformity in texture, the permeability of small molecules across a thin cast film of P-50 was observed, and compared
with the case of an oriented film. It is shown in Table 8 that the
permeability reduces by one order of magnitude only by 50% stretch
[41]. Inside the fiber, the molecules are much more extended and
oriented through the superdraw. Therefore, migration of small molecules and ions into the fiber must be so much retarded that it ensures apparent chemical resistance of the fiber.

4.5 Applications

Through various different after treatment, the copolymer aramid
fiber Technora is finding use in almost all areas where crystalline
PPTA fibers have been applied, except for some special uses. For

Table 1.8 Permeability of Small Molecules across P-50 Membranes[a]

	Membrane	
	As cast	Drawn ($\times$ 1.5)
Thickness (mm)	0.029	0.020
Density (g/ml)	1.3375	1.3755
Permeate gas	Permeability (ml·cm/cm^2·sec·mmHg)	
H_2O	3.7×10^{-11}	4.0×10^{-12}
O_2	1.5×10^{-12}	1.4×10^{-13}
N_2	2.6×10^{-13}	3.9×10^{-14}

[a]Method: JIS Z1707 (1975).

volume application in tires, its improved fatigue resistance [33] is favored. Various hoses are other areas where the copolymer fiber is appreciated due to its chemical and thermal stability. Uses where its inherent feature is fully appreciated are the reinforcement of cement concrete in a form of chopped fibers both for ordinary and autoclave cure types [42]. A new technique of fiber processing is applied to give a spannized yarn, which is expected to be used for making a protective garment [43].

ACKNOWLEDGMENT

The authors wish to express their thanks to their colleagues at Product Development Research Laboratories for their contributions to the research work reported in this chapter, particularly to K. Shimada, H. Mera, A. Aoki, T. Nishihara, Y. Nakagawa, and T. Yamada. They also acknowledge the support and encouragement of the management of Teijin Limited, especially H. Itagaki and T. Fukushima. Finally, the authors appreciate the help and understanding extended by T. Sasaki and S. Hiratsuka.

REFERENCES

1. E. I. du Pont de Nemours & Co., Catalog, 1972.
2. T. Kajiyama, *Kobunshi/High Polymers, Japan*, *32*, 336 (1983).
3. G. S. Fielding-Russell, *Text. Res. J.*, *14*, 861 (1971).
4. S. L. Kwolek (for du Pont), U.S. Patent 3,600,350, August 17, 1971.

5. T. Bair and P. W. Morgan (for du Pont), U.S. Patent 3,817,941, June 18, 1974.
6. H. Blades (for du Pont), U.S. Patent 3,767,756, October 23, 1973.
7. M. Jaffe and S. Jones, in *High Technology Fibers, Part A* (M. Lewin and J. Preston, Eds.), Marcel Dekker, New York, 1985, p. 349.
8. M. G. Dobb, D. J. Johnson, and B. P. Saville, *Polymer, 22*, 960 (1981).
9. S. L. Phoenix and J. Skelton, *Text. Res. J., 44*, 934 (1974).
10. K. Yabuki, H. Ito, and T. Ohta, *Sen-i Gakkaishi, 31*, T-524 (1975); ibid., *32*, T-55 (1976).
11. M. G. Dobb, D. J. Johnson, and B. P. Saville, *J. Polym. Sci., Polym. Phys. Ed., 15*, 2201 (1977).
12. S. Manabe, S. Kajita, and K. Kamide, *Sen-i Kikai Gakkaishi, 33*, T-93 (1980).
13. M. Panar, P. Avakian, R. C. Blume, K. H. Gardner, T. D. Gierke, and H. H. Yang, *J. Polym. Sci., Polym. Phys. Ed., 21*, 1955 (1983).
14. K. Haraguchi, T. Kajiyama, and M. Takayanagi, *J. Appl. Polym. Sci., 23*, 903, 915 (1979).
15. Asahi Chem. Ind., Japanese Patent (disclosed) 59-1710, January 7, 1984.
16. K. Matsuda, *ACS Polym. Preprints, 20(1)*, 122 (1979).
17. S. Ozawa, *Preprints for 82/2 Lecture Series, Soc. Polymer Sci., Japan*, October 1982, p. 30.
18. H. Mera, T. Nishihara, and A. Aoki, unpublished.
19. Unitika Co., Japanese Patent (disclosed) 49-62720, June 18, 1974.
20. Y. Nakagawa, H. Mera, and T. Noma, unpublished.
21. K. Matsuda, T. Shinoki, and A. Aoki, unpublished.
22. S. Ozawa, K. Matsuda, Y. Nakagawa, and T. Nishihara, unpublished.
23. Teijin Co., Japanese Patents (disclosed) 51-76386, 51-134743, 51-136916, and 52-98795 and U.S. Patent 4,075,172, prior. December 27, 1974.
24. Bayer A.G., Japanese Patent (disclosed) 48-35117, May 22, 1973.
25. L. B. Sokolov, V. D. Gerasimov, V. M. Savinov, and V. K. Belyakov, *Thermally Stable Aromatic Polyamides* (in Russian), Izdatelstov, Ximiya, Moscow, 1975, chap. III.
26. K. Matsuda, unpublished.
27. Teijin Co., Japanese Patent (disclosed) 53-33294, March 29, 1978.
28. P. Smith and P. J. Lemstra, *J. Mater. Sci., 15*, 505 (1980).
29. Monsanto Co., Japanese Patent (disclosed) 53-58023, May 25, 1978.

30. T. Jinda and T. Kawai, *Sen-i Gakkaishi*, *37*, T-279 (1981).
31. S. Ozawa, *Polym. J.*, *Japan*, *19*, 119 (1987).
32. K. Kazama, *Preprints for Plastics Engineering Forum*, *Soc. Polymer Sci.*, *Japan*, March 1981, p. 3.
33. M. Kamiyoshi, *Plastics*, *Japan*, *36(3)*, 26 (1985).
34. T. Takata, *Zairyo Kagaku*, *21*, 348 (1985).
35. M. Kamiyoshi, unpublished.
36. R. J. Morgan, C. O. Pruneda, and W. J. Steele, *J. Polym. Sci.*, *Polym. Phys. Ed.*, *21*, 1757 (1983).
37. M. Horio, T. Kaneda, S. Ishikawa, and K. Shimamura, *Sen-i Gakkaishi*, *40*, T285 (1984).
38. K. Matsuda, T. Nishihara, and T. Watanabe, *1984 Int. Chem. Congress of Pacific Basin*, 09P02.
39. K. Matsuda, unpublished.
40. A. Aoki and T. Takata, unpublished.
41. T. Yamada, unpublished.
42. S. Akihama, H. Nakagawa, T. Tada, and M. Yamaguchi, *3rd Int. Symp. on Development in Fiber Reinforced Cement and Concrete*, July 1986.
43. A. Takagi, *Kasen Geppo*, *No. 2*, 64 (1986).

2

ARAMID STRUCTURE/PROPERTY RELATIONSHIPS AND THEIR ROLE IN APPLICATIONS DEVELOPMENT

DAVID TANNER, JAMES A. FITZGERALD, and PAUL G. RIEWALD /
E. I. du Pont de Nemours & Company, Inc., Wilmington, Deleware

WARREN F. KNOFF / E. I. du Pont de Nemours & Company, Inc.,
Richmond, Virginia

1. INTRODUCTION

Aramid is a generic term used to describe fibers from wholly aromatic polyamides. Since about 1965, several types of aramid fibers have been commercialized or are under development. Examples include

35

fibers from poly(m-phenyleneisophthalamide) [MPD-I], poly(p-phenyleneterephthalamide) [PPD-T], and a copoly-terephthalamide from p-phenylene diamine and 3,4'-diaminodiphenyl ether [PPD/3,4'POP-T]. Figure 1 shows the chemical structures of these polymers.

The meta-oriented aramids are used where outstanding thermal and electrical properties are required, as in thermal protective apparel and electrical insulation. The para aramids and the copolymer fibers are capable of very high strength and modulus and are used, for example, in reinforced rubber and plastic applications.

The scope of this chapter will include para-aramids (p-aramids) that are formed from anisotropic or liquid crystalline solutions, for example, PPD-T. The scientific breakthrough leading to high performance fibers from liquid crystalline solutions and the key technology elements have been reported previously by many authors [1,2,18]. This chapter provides an overview of this technology with emphasis on the fundamental relationship between polymer molecular structure and the physical structure and properties of the fibers. Then the translation of this scientific discovery into

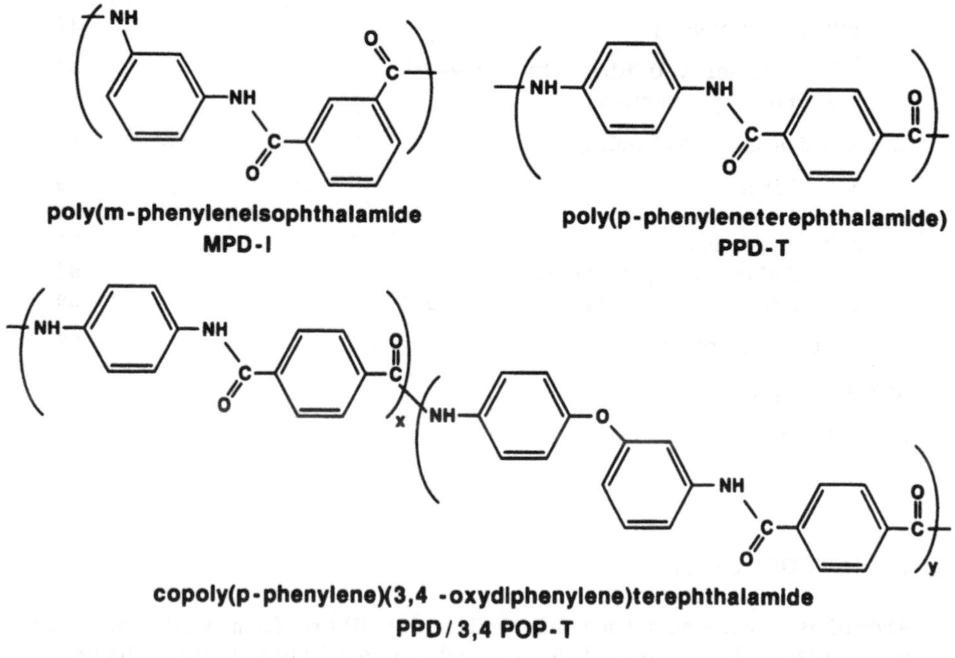

poly(m-phenyleneisophthalamide
MPD-I

poly(p-phenyleneterephthalamide)
PPD-T

copoly(p-phenylene)(3,4-oxydiphenylene)terephthalamide
PPD/3,4 POP-T

Figure 2.1 Polymer structures of some commercial or developmental aramid fibers.

practical industrial applications is described. Finally, through several examples, future research directions in this area of technology are mentioned.

2. FIBER TECHNOLOGY

2.1 Polymer and Fiber Structures

Poly(para-phenylene terephthalamide) is a rigid molecule (Fig. 2). The rotation around the para-positioned bonds on the aromatic ring introduces only a limited amount of flexibility. In addition, the carbon-nitrogen bond of the amide group has considerable double bond character, which severely restricts rotation about that bond. This partial double bond leads to a resonance-conjugated system. One result of this is that the fiber from PPD-T is yellow while MPD-I, a polymer in which the linkages are meta so that conjugation through the amide bonds cannot exist, is white. The rotational activation energy of the carbon-nitrogen bond is approximately 22 kcal/mol; whereas if it was purely a single bond, the rotational activation energy would be 2-4 kcal/mol. Overall, the restricted rotation contributes greatly to the stiffness of the molecule. This rigidity is a key property utilized in the manufacturing of p-aramid fibers.

The behavior of rigid molecules in concentrated solutions is very different from that of flexible coil-type molecules and is an important property allowing fiber production. Rigid molecules can move relatively independently in very dilute solutions, but their movement and ability to pack together randomly becomes quite restricted at higher concentrations. At a critical concentration a transition to a liquid crystalline state often occurs [3,4]. In this state the molecules are aligned parallel to each other in regions that are randomly oriented with respect to each other. This is quite different from flexible molecules, which pack together at higher concentrations by bending and entangling.

The behavior of the liquid crystalline solution under shear and elongation is a critical property that allows the production of high strength, high modulus, as-spun fibers. At rest, the molecular orientation of the ordered domains is random with respect to each other. Under shear, the liquid crystalline domains become elongated and oriented in the direction of the deformation [5-7].

In the fiber manufacturing process the unoriented spin solution is fed into the spinneret, where partial orientation and elongation take place. When solution exits the spinneret capillary, a partial deorientation takes place due to the elastic properties of the solution. The solution is then attenuated, and a high degree of orientation

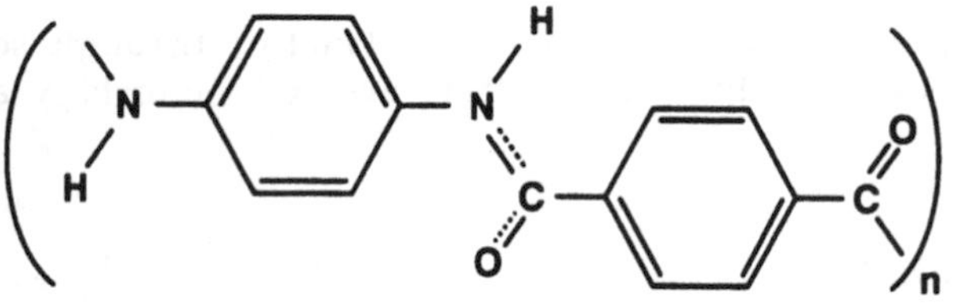

Paraposition bonds
• High frequency, inherently rigid

Carbon-nitrogen bond
• Double bond character, limited rotation

Figure 2.2 Structural formula of the PPD-T repeat unit showing resonance stabilization.

is attained. The high orientation achieved in the fiber is then captured via quick cooling followed by coagulation with water [8,9].

PPD-T fibers are highly crystalline materials. Figure 3 shows a schematic of the unit crystal cell of PPD-T, which has been determined via wide-angle X-ray experiments [10]. The molecules are arranged in parallel hydrogen-bonded sheets. Two of them are illustrated on the front face of the unit cell. There will be another set of molecules on the back face in exactly the same arrangement, and there will be another sheet that will pass through the center of the unit cell. That can be seen more clearly from the top view looking down the molecular axis of the fiber. The benzene rings in the polymer chain are canted alternately for maximum packing efficiency. The theoretical density of the unit cell is 1.48; the observed fiber density is 1.45 g/cm^3.

Another aspect of the crystallinity of some p-aramid fibers is lateral crystalline order [11,12]. Figure 4 shows three types of lateral or transverse crystalline arrangements that could exist in a fiber. The upper fiber cross section is representative of a fiber that has a random crystal orientation. The center one is a fiber that has a radial crystal orientation, and the lower one is a fiber that has tangential crystalline orientation. p-Aramid fibers can have a well-defined radial crystalline orientation that is characteristic only of aromatic polyamide fibers spun using a dry jet-wet spinning process. This type of radial order had never been observed before in a synthetic organic fiber. All other highly oriented fibers

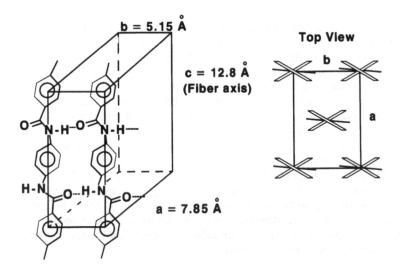

Figure 2.3 The unit crystal cell of PPD-T, as determined by wide-angle X-ray scattering of the fiber.

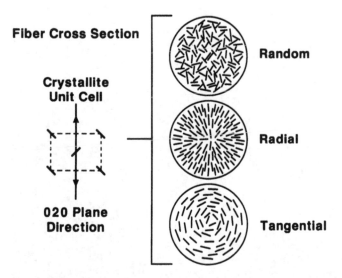

Figure 2.4 Types of transverse or lateral order that can occur in fibers, viewed looking down the fiber axis.

studied have had the random lateral order shown in the top schematic. The very existence of large scale, three dimensional crystalline
order in a fiber is highly unusual, if not unique.

Although first observed and quantified using optical birefrigence
and electron diffraction techniques, the lateral crystalline order
has been most dramatically demonstrated with electron transmission
photomicrographs of silver sulfide-impregnated thin sections of
the fibers (Fig. 5) [13].

The very obvious fibrillar nature of p-aramid fibers is also
an aspect of the structure that influences the fiber properties.
Figure 6 shows the highly fibrillated broken ends of PPD-T fiber
from a loop break and tensile failure. It is highly possible that
the fibrils are the coagulated and solidified residue of the elongated
oriented liquid crystalline domains prior to quench. The domain
ends, then, are likely to be the structural defect presently limiting
the fiber tenacity (see next section).

Figure 2.5 Radial morphology of a p-aramid fiber. (From Ref. 13.)

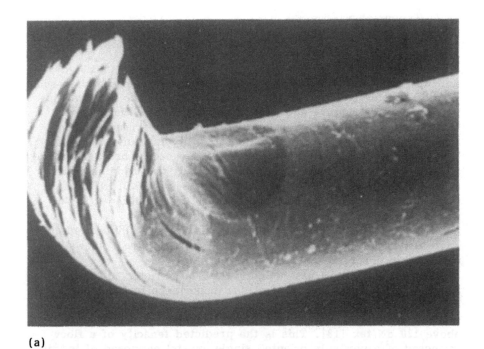

(a)

(b)

Figure 2.6 Broken ends of PPD-T fibers from (a) a loop break and (b) tensile failure.

2.2 Fiber Properties

End-use applications utilize the mechanical properties of a p-aramid fiber. Therefore, a fundamental understanding of the relationships between fiber structure and mechanical properties is important. The tensile modulus or extensional stiffness of p-aramid is a key mechanical property determining its suitability in many end uses. While the theoretical modulus of the PPD-T molecule has been calculated to be greater than 1500 dN/tex [14-16], the modulus of commercially available p-aramid products ranges from 440 dN/tex for fiber dried under low tension and temperature conditions, to 900 dN/tex or more for fibers heat treated under tension.

Fiber modulus is a strong function of the crystalline orientation angle determined by wide-angle X-ray [1,17]. Figure 7 shows the relationship observed for one set of samples.

The strength on a unit weight basis, or tenacity, of p-aramid fibers is a property that is of critical importance in many end uses, but it is also one of the most difficult to understand. The theoretical tenacity of p-aramid fibers has been estimated to be above 120 dN/tex [18]. This is the predicted tenacity of a fiber composed of a perfectly oriented single crystal composed of infinitely long polymer molecules. Yarn tenacity of a typical p-aramid yarn is 21 dN/tex, although presently, p-aramid yarns having a tenacity of 25 dN/tex have been supplied to the aerospace industry. There is still much room to enhance the tenacity to fully capitalize on the basic structure of the PPD-T molecules.

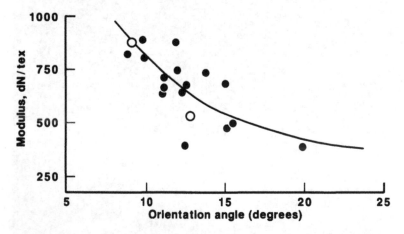

Figure 2.7 Plot of PPD-T fiber modulus versus crystalline orientation angle. (From Ref. 1.)

There are several aspects of p-aramid fibers that are responsible
for their high tenacity: molecular weight, orientation, fiber structure,
and the presence (or absence) of flaws such as particulates or voids
that may represent stress concentration points. The high molecular
weight of the PPD-T used to manufacture p-aramid fiber (M_n greater
than 20,000) is important in attaining high strength. Electron para-
magnetic resonance studies suggest that bond rupture is the predominant occurrence during failure [19]. This indicates that failure
of molecules is the major fiber tensile failure mechanism rather than
pullout of molecules away from neighbors, associated with lower
molecular weight.

Early in the development of high tenacity organic fibers, it
was recognized that high tenacity also required high levels of molecular orientation. This is necessary so that the high covalent bond
strength can be realized in the fiber axis direction. Present p-aramid
products have the needed very high molecular orientation (less
than 12°).

The tenacity/orientation relationship is indicated in Fig. 8,
where tenacity is plotted against modulus (refer to Fig. 7). The
correlation appears most strong at moduli corresponding to orientation
angles greater than about 12-19°. Below that, scatter increases

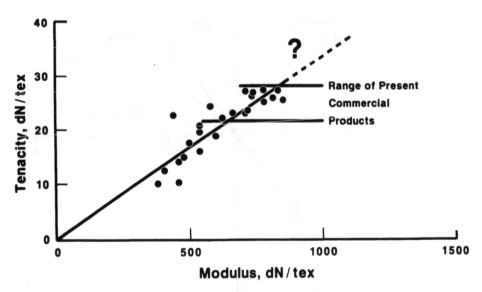

Figure 2.8 Plot of PPD-T fiber tenacity versus modulus. Modulus
is an inverse function of orientation angle (see Fig. 7).

and it appears that the fiber tenacity is not able to capitalize on improved orientation, so that another mechanism begins to dominate tensile failure.

The tensile failure of unfatigued p-aramid fibers generally occurs over lengths of greater than 50× the fiber diameter, and fibrillation is often observed as shown in Fig. 6. Assuming that fiber failure initiates at the fibril ends and propagates via shear failure between the fibrils, the fiber tensile strength will be strongly influenced by the fiber axial shear modulus and strength.

Another structural feature that is important for obtaining high tenacity fibers is the absence of defects. This can include particulates, voids, cracks, and regions of high concentration of molecular ends.

The compressional properties of p-aramid fibers are substantially different from the tensile properties, as shown in Fig. 9. While in tension, the stress/strain behavior is very close to elastic; in compression, there is a yield at approximately 0.4% compressional strain. Physically, the compressional yield corresponds to the formation of what are often referred to as "kink bands."

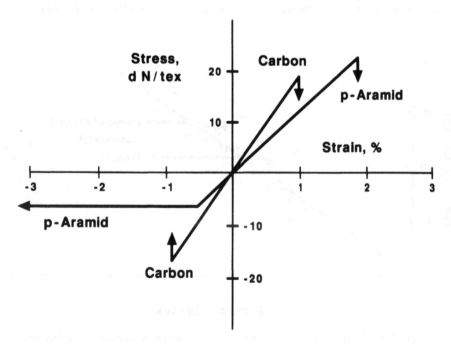

Figure 2.9 Stress-strain curve of PPD-T fiber, showing tension in the NE quadrant and compression in the SW quadrant.

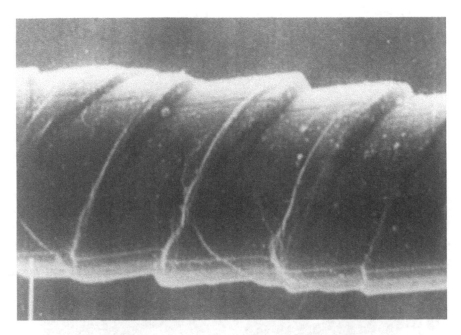

Figure 2.10   Severely link-banded fibers. (From Ref. 20.)

Figure 10 shows a photograph of a fiber that has been severely compressed and has extensive kink bands. These bands appear as disruptions in the structure occurring at 50-60° to the fiber axis.

On a molecular scale, the formation of kink bands is thought to correspond to a molecular rotation of the amide-carbon-nitrogen bond from the normal extended "trans" configuration to a kinked "cis" configuration as shown in Fig. 11. These rotations must occur in pairs, and propagate across the unit cell, across fibrils, and finally, across the entire fiber. Figure 12 shows the tensile failure of a severely kink-banded fiber [20]. Note the sharp angular break as opposed to normal tensile failure in Fig. 6.

The flexural and bending properties of the fibers that result in the textilelike nature of p-aramid yarns are functions of both the small filament diameter and the compressional yield. The bending moment of p-aramid fibers is comparable to that of continuous-filament industrial nylon. The low critical radius of curvature, the point at which the compressional strain on the inside of a bent fiber equals the critical compressive strain of the material, is about 1.5 mm and allows p-aramid to be handled on most textile processing

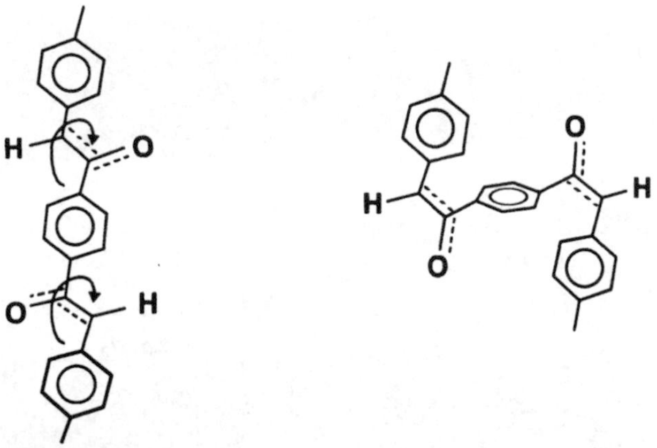

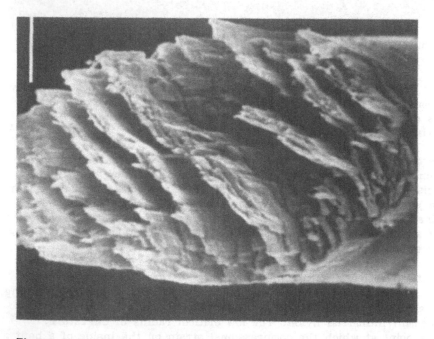

Trans   **Cis**

Figure 2.11 Molecular buckling of PPD-T under compressional strain.

Figure 2.12 Photograph of tensile failure of a severely kink-banded fiber. (From Ref. 20.)

machinery. The above properties also result in high energy-absorbing ability of p-aramid and its excellent ballistic performance.

Fibers from PPD-T do not have a literal melting point or even a glass transition temperature as is normally observed with other synthetic polymers. In spite of this, however, the fibers have a remarkable response to heat treatment under tension [21]. Unlike most other as-spun fibers, drawing is not possible with an aramid fiber prepared by a dry jet-wet spun process. At most, an extension of less than 5% is possible, even with temperatures exceeding 500°C. Under these conditions, aramid fibers achieve an improvement in orientation (from 12-15° to about 9° or less) and an increase in crystalline perfection and crystallite size. Most striking is the increase in modulus, from about 500 dN/tex to over 900 dN/tex. The stress-strain curves of the two forms of aramid fibers are shown in Fig. 13.

To summarize, it has been possible to capitalize on the rigidity of the PPD-T molecule by spinning liquid crystalline solutions to generate fibers with unusual properties. The crystallinity and orientation of the solution are translated to the fiber.

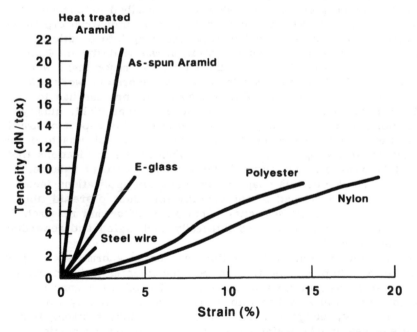

Figure 2.13 Stress versus strain for several industrial fibers; stress as dN/tex, strain as percent extension based on original sample test length.

The aromatic structure confers excellent thermal properties, including a zero strength temperature over 600°C. The organic polymer base results in low density. The polyamide structure confers thermal and chemical stability. The para orientation provides the ultra high crystallinity via extended chain registry and hydrogen bonding as well as the propensity toward a lyotropic structure. The dry jet-wet spinning process, through extensional shear of the proper high-density liquid crystalline spinning solution, perfects the molecular orientation and creates the unique three-dimensional crystalline order present in these fibers. The order present in the fibers, then, provides the tenacity, modulus, and other properties that are creating the excellent end-use potential being observed today. The very fine denier per filament resulting from the air-gap attenuation provides textile processibility and flex fatigue resistance.

3. APPLICATIONS TECHNOLOGY

The discovery of a fiber with a new chemical composition and with an entirely novel balance of properties is a highly unusual event. The elements of the underlying technology have been described earlier in this chapter and have also been described extensively elsewhere [1,2,5-7]. The question that faces an industrial organization now is "How can we translate this discovery into practical, high value applications?" When the decision was made by du Pont to drive forward with the aramid discovery, two major tasks were undertaken. The first was translation of the laboratory work into the elements of a commercializable process. This process had to meet the requirements of cost and investment, fiber yield, quality, and uniformity to achieve a profitable venture. This aspect of aramid development will not be covered here. The second task was to develop the end-use technology to use the new fiber in the applications that would take maximum advantage of the available property balance.

The remainder of this chapter describes the applications technology that was developed to successfully introduce p-aramid fibers into a variety of market segments. This phase of aramid research and development has been as scientifically challenging and rewarding as the discovery itself.

Market development of a new fiber can be an enormously complex task. Use of a high strength, high modulus fiber as a direct replacement for an incumbent fiber in an industrial application generally does not work. Unless taken into consideration, secondary properties beyond tenacity, elongation, and modulus often play a strong role in the end-use article performance. These secondary properties include compressive strength, abrasion resistance, fiber surface chemical reactivity, and temperature dependence of tensile properties.

In order to maximize the value of a new fiber in an application, technology must be developed for the highest possible performance of the final product, e.g., a tire or a fiber-reinforced plastic article. Two approaches must be used in the development of the end-use technology. First, a fundamental understanding of the structure of the fiber must be obtained and an understanding of how the fiber structure affects the tensile and secondary properties. This is important in the design and performance of the fabricated article. Second, the fiber and all the rest of the components in the article must be considered together in the design. In other words, a *systems approach* must be used to take full advantage of the desirable properties of the fiber in order to design around negative features of the fiber as well as to capitalize on any possible synergistic combinations of properties of the other components in the system.

The next section describes several industrial applications, including tires, ropes and cables, ballistics, asbestos replacement, and composites.

3.1 Tires

3.1.1 Introduction

p-Aramid fiber has particular utility in reinforcement of radial tires, both in the belt, where modulus contributes to tire performance, and in the carcass, where strength contributes to tire durability.

The key needs of a tire system are impact and fatigue resistance, tire life and durability, tread life, cornering and handling capability, and high speed performance. The required reinforcing fiber properties are high strength and modulus, strength retention after fatigue, and adhesion to rubber [22].

p-Aramid fiber excels in these applications versus steel, the incumbent reinforcing material (Table 1). Note particularly the advantages in tread wear, weight, running temperature, and fatigue in truck tire carcasses, and ride, weight, and durability in passenger car tires.

A combination of a systems approach and knowledge of fiber structure and properties was necessary to realize this successful application.

3.1.2 Fatigue Performance

A key concern in the design of the passenger car tire belt reinforcement package has been fatigue performance [23]. The major mechanism for strength loss of an aramid fiber such as p-aramid was determined to be compression fatigue. If an aramid fiber is subject to compressive strain, kink bands develop, as described earlier, and repeated or intense compression can lead to strength loss.

Table 2.1 Comparison of p-Aramid Versus Steel in Radial Tires

Truck tires—Carcass[a]	
Lower treadwear	10%
Lighter weight	5%
Cooler running	5°C
Improved fatigue resistance	2×
Passenger car tires—Belt[b]	
Softer ride	Yes
Lighter weight	20%
Higher durability	Yes
High speed performance	Equal
Treadwear	Equal

[a]Truck tires built by tire manufacturers for comparison.
[b]Passenger car tires built by du Pont for comparison.

Figure 14 shows the results of subjecting filaments to compressive strains of 0.5%, 0.8%, and 1.0%. In all cases, kink bands appeared by 10 cycles. At a strain of 0.5%, no strength loss was observed out to several hundred thousand cycles. At higher strain levels, strength loss appears at 1000 to 10,000 cycles and progresses with increasing number of cycles.

It is clear, then, that cyclic axial compressive strain on filaments in a tire cord must be held to less than about 0.5-0.8%. There are at least two ways to accomplish this: tire cord construction and placement of the reinforcing cords.

Tire cords are prepared by twisting yarns or filament bundles, then combining the twisted yarns together into a cord by twisting in the opposite direction. Putting fibers together in this way has been done for many years with most fibers that have been used for tire reinforcement [22]. In a twisted structure, the matrix freedom of the filaments provides improved fatigue resistance, since the filaments can move to relieve the stress. This ability to reduce compressive strains on individual filaments increases with increasing twist level. There is a trade-off, however. Tenacity and modulus decrease with increasing twist, since at higher twist levels the helix angle between the filaments and the cord axis provides less and less axial strain on the filament as axial strain is placed on the cord. Since the transverse properties of these highly oriented fibers are on the order of 50% or less of the axial properties, there is a strong dependence of both tenacity and modulus on helix angle and hence on the cord twist level (tenacity varies as $\cos^2$, modulus varies as $\cos^4$) [24]. As a result, a balance must be struck to

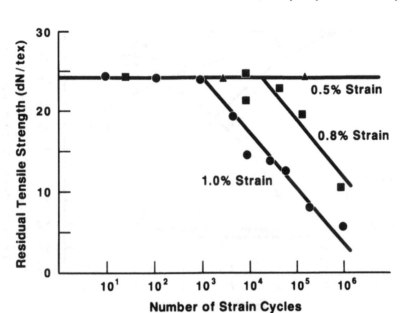

Figure 2.14   Percent strength retention as a function of the number of compression cycles at 0.5%, 0.8%, and 1.0% axial compressive strain.

provide the optimum set of properties for a given end use. Figure 15 illustrates this. Twist multiplier, at a constant cord denier, is a direct indication of the filament helix angle [24].

Two routes now are evident to provide a tire design that minimizes fatigue strength loss while maintaining desirable cord tensile properties to provide satisfactory tire performance. The first is optimization of cord twist level for fatigue versus tenacity and modulus. The second is to provide selective cord placement in the belt to reduce individual filament compressive strain in critical zones by having more cords in this zone to share the overall compressive stresses, for example, during cornering—especially at the belt edge.

It is possible to design the reinforcement package to minimize compressive strains and obtain excellent fatigue performance in passenger car tire belts.

As mentioned earlier, strength loss in cords in a tire belt occurs predominantly at the belt edge due to compressive forces during cornering. Using tires built for comparison, a 38,000-km fleet test was run (Fig. 16), and as expected, confirmed that high twist multiplier decreases filament compressive strain and thus improves fatigue performance. Use of folded belts, an example of selective

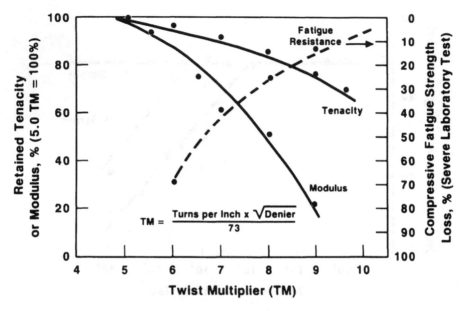

Figure 2.15 The relationship among cord tenacity, modulus, and fatigue with cord twist multiplier. A proposed optimum would be between twist multipliers 6 and 7.

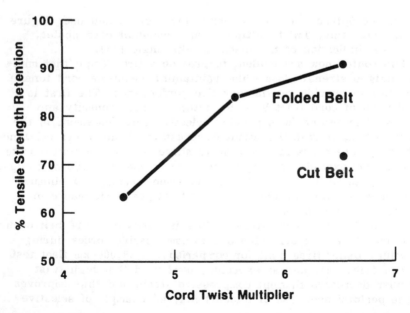

Figure 2.16 Percent strength retention of cords removed from belt edge after fleet test, as a function of cord twist multiplier.

cord placement, places more reinforcement at the belt edge, which correspondingly decreases compressive strain and improves fatigue performance. Utilizing these two routes to improved fatigue perform- ance, fully satisfactory strength retention can be obtained with as little as 1 lb of aramid fiber replacing 5 lb of steel wire.

In addition, comparison of these various belt designs, particu- larly the lower twist level folded belt design versus the traditional cut edge design, provided a number of other advantages: improved ride and handling, better endurance, lower rolling resistance (related to fuel economy), and superior high speed performance.

3.1.3 Cord-to-Rubber Adhesion

Adhesion of a reinforcing fiber to rubber is an essential criterion for acceptability of a tire cord condidate. Adhesion technology for cotton, rayon, nylon, polyester, glass, and steel has been developed to the point where the weakest link in the cord (fiber)/adhesive/ rubber interface is in fact the rubber itself. In tests for adhesion, if the stress required to separate a reinforcing fiber from a matrix results in failure of the matrix, then fully satisfactory adhesion is clearly present.

Nylon 6,6 has the same functional groups as aramid polymer. Adhesion of 6,6 nylon to rubber is easily achieved by dipping the tire cord in a suspension of a resorcinal-formaldehyde latex (RFL) mixture in water and then curing at about 250°C. The RFL easily bonds to the nylon 6,6 substrate, and even after curing, is suffi- ciently reactive to chemically bond to rubber during the normal rubber curing process. Unlike 6,6 nylon and in spite of the similar- ity of chemical functionality, p-aramid fibers are much more difficult to bond to rubber. The problem very likely is due to (a) the high crystallinity of the p-aramid fiber, which creates difficulty in gaining chemical access to the amide groups, and (b) decreased reactivity of the amide group caused by the strength of the intercrystalline hydrogen bonds, as well as the resonance stabilization of the amide bond within the semiconjugated system.

The adhesion system for p-aramid fibers used today by several tire manufacturers involves treating a twisted cord of aramid with an epoxy resin followed by a second coating of an RFL emulsion. This process provides a strong bond between the aramid fiber sur- face and the epoxy layer, a strong bond between the epoxy layer and the RFL layer, and finally, a strong bond between the RFL layer and the rubber matrix. Table 2 illustrates typical formulations and treating conditions for tire cord adhesion processing. Figure 17 shows a schematic of an adhesion system used with aramid fibers, and the particular interfaces where shear forces may cause failure— and where the chemistry becomes very important.

Table 2.2 Preferred Conditions for Adhesion Treatment of Aramid
Tire Cords

	Oven 1	Oven 2
Oven conditions		
Dip applied	Subcoat	Topcoat
Oven temperature, °C	243	232
Residence time, sec	60	60
Oven exit tension, dN/tex	0.88	0.26

	Percent by weight
Subcoat composition	
Diglycidyl ether of glycerol[a]	2.2
10% NaOH	0.28
5% "Aerosol" OT	0.56
Water	96.94
Topcoat composition	
28% ammonium hydroxide	1.1
75% RF resin	3.9
41% vinylpyridine latex	43.7
37% formaldehyde	2.0
33% wax dispersion[b]	2.9
25% carbon black dispersion	10.8
Water	35.6

[a]This is an epoxy compound. Many compounds of this class are
skin irritants, and several have been shown to be mutagenic in
bacterial tests. There is some evidence of increased tumor incidence
in animal feeding studies with some epoxy compounds, but we know
of no evidence relating these materials with human carcinogenesis.
Instructions on safe handling of dip ingredients are customarily
provided by the vendor. In the absence of specific instruction,
we recommend that precautions be taken to completely avoid skin
contact and inhalation of epoxy compounds in preparing and applying dips to p-aramid yarns.
[b]"Acrawax" C is claimed as an additive to dips in U.S. Patent
3,876,457, April 1975, issued to Uniroyal.

Fiber Embedded in a Rubber Matrix

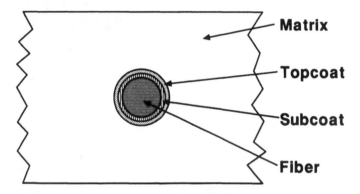

Interfaces: Matrix/Topcoat
Topcoat/Subcoat
Subcoat/Fiber

Figure 2.17 Schematic representations of the chemical systems in a cured-in-rubber tire cord.

There has been continuing development of novel belt reinforcement concepts such as Tredloc®, which is a woven arrangement of cord that avoids placing cord ends at the belt edge that provide stress concentration points. We expect that this type of research will continue.

There is much discussion about the role of the cast or injection molded tire in the future. Incorporation of aramid fiber reinforcement in a cast tire would undoubtedly improve its performance, but considerable work will need to be done to optimize use of aramid fiber for reinforcement of this tire.

Approaches similar to the radial tire application have been taken in development of technology to utilize p-aramid in mechanical rubber goods such as conveyor belts, power transmission belts, and automotive and high pressure hydraulic hoses.

3.2 Ropes and Cables
3.2.1 Introduction

Another important application for p-aramid is ropes and cables. The high specific strength and inherent low elongation are the major driving forces for use of p-aramid fibers in ropes and cables.

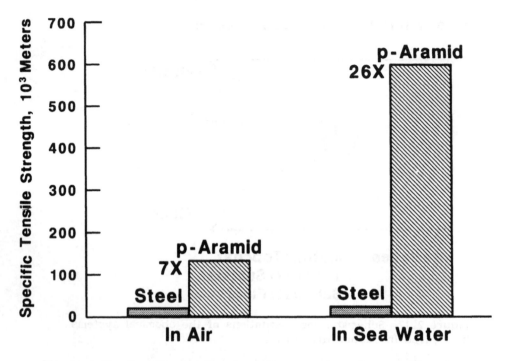

Figure 2.18   Specific tensile strength comparison for aramid and steel cables, in air and in water.

The strength-to-weight ratio of p-aramid compared with steel wire is shown in Fig. 18.

The specific strength is calculated by dividing tensile strength by weight per unit length and can be interpreted as the length of a strength member that will break under its own weight. p-Aramid has a much higher specific strength than steel; the comparison is particularly favorable in seawater, where p-aramid specific strength is more than 20 times that of steel. This advantage offers potential for smaller, lighter, more easily handled ropes and cables; and, in long lengths where the self-weight of steel becomes critical, p-aramid can offer more payload, higher safety factors, or the possibility of performing missions previously not feasible. Other properties that are important advantages for p-aramid in ropes and cables are excellent tension-tension fatigue resistance, corrosion resistance, low creep, nonconductivity, and excellent cycling-over-pulley life [25].

A wide variety of rope constructions (Fig. 19) has been demonstrated from p-aramid fibers, including three-strand, eight-strand plaited, parallel lay, braids, and low twist, multistrand ropes of wire rope-type constructions. In properly made ropes of comparable construction, those made from p-aramid will have about twice the strength of nylon or polyester ropes at equal weight and about the same strength as steel wire rope, but at 20-25% the weight.

Extensive testing has shown that multistrand ropes of p-aramid similar to steel wire ropes have excellent strength translation and good durability as working ropes, and hence they are the preferred constructions for many applications. However, for each major use, construction designs must be selected and the technology developed to capitalize on the unique features of p-aramid while minimizing negative features such as poor abrasion resistance and low damage tolerance.

3.2.2 Fatigue Performance

An example of an engineered rope structure is a riser-tensioner line used on floating offshore oil drilling platforms to keep the riser pipe at a constant elevation and under uniform tension while the platform surges with the waves (Fig. 20). These 44-mm-diameter

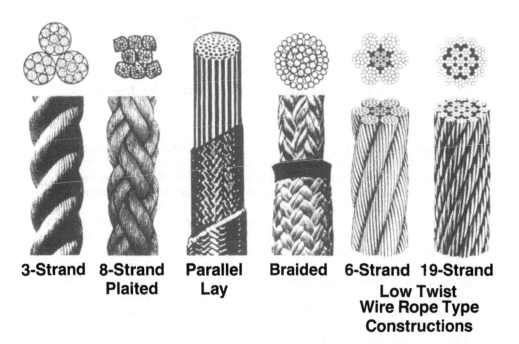

3-Strand   8-Strand   Parallel   Braided   6-Strand 19-Strand
Plaited   Lay   Low Twist
Wire Rope Type
Constructions

Figure 2.19 Rope constructions made from p-aramid.

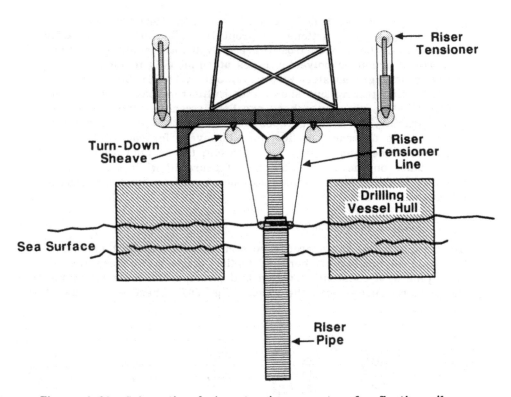

Figure 1.20 Schematic of riser tensioner system for floating oil platforms.

ropes see considerable cycling over pulleys, also called sheaves [26,27]. Previous laboratory studies had shown that small-diameter ropes of p-aramid could far surpass steel in cycling performance, but scale-up of the best small constructions to 44 mm diameter gave rope lifetimes only 5-10% that of steel wire rope.

Analysis showed that internal loads in twisted rope rise rapidly with increasing diameter, which leads to high frictional heating, high internal abrasion, and shear fatigue failure of yarn as rope elements move by each other when the rope is bent [28,29]. Modest abrasion and shear properties of p-aramid, related to its fibrillar structure, lead to failure under these high stress conditions. As shown in Fig. 21, these internal loads stem from radial squeezing forces that also increase rapidly with increasing twist, bearing pressure against the pulleys and bending stresses.

During the course of the riser tensioner application development, several design changes were made that improve rope lifetime over 50-fold (Fig. 22). These included using 18-strand ropes with the strands nested in ways that minimize crossovers, thereby avoiding concentrated loads (5× improvement); jacketing each strand with a braid impregnated with fluorocarbon resin and silicone oil to reduce friction, thereby reducing heating, abrasion and internal shear stresses (5.9× improvement); and reducing twist to reduce radial forces (1.7× improvement).

The technology developed for p-aramid use in riser tensioners resulted in a rope having more than three times the life of steel in severe laboratory tests and more than five times in service.

3.2.3 Torque-Tension

A future application under development is deep-water mooring lines for the same type of drilling platforms that use riser-tensioner lines.

Most oil rigs use steel chain mooring systems, which are limited to water depths less than 450 m because the high sag of the heavy steel chain reduces the restoring force required to hold the platform in position [30,31]. This restoring force is proportional to the horizontal component of line tension shown as H in Fig. 23. Because

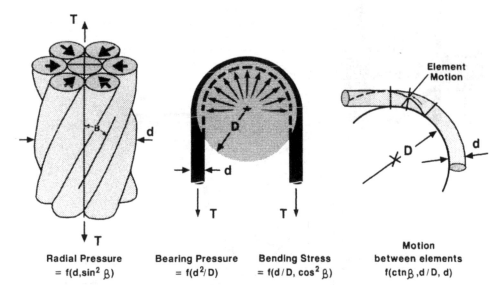

Figure 2.21 Internal forces and motions in ropes.

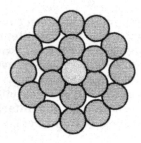 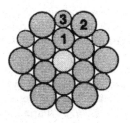

Former Construction

- **Strands same size**
- **Unlubricated**
- **Layers at same helix angle**
- **High helix angle**

Redesigned Rope

- **Three strand sizes**
- **Lubricated (5.9x)**
- **Strands nested (5.x)**
- **Lower helix angle (1.7x)**

Figure 2.22   Design changes for improved riser tensioner performance.

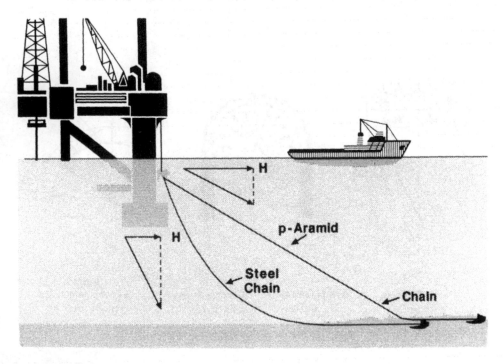

Figure 2.23   p-Aramid for deep-water oil platform mooring.

p-aramid is lighter, sag is less and restoring force is substantially greater. At 730 m depth, the restoring force with p-aramid mooring lines is 80% higher than that of steel chain.

Calculations indicate that lines of p-aramid could perform at water depths up to 3000 m or deeper. This capability allows design of vessels for very deep water use and a way for existing platforms to extend their depth capability by insertion of p-aramid between chain or steel wire rope.

Simple substitution is not that easy, however, since all lines are part of the total mooring system and must be compatible. For instance, a particular system being considered requires that a p-aramid rope of 450 tons break strength be connected to a steel wire rope, which, because of its constructional geometry, generates a high torque when tensioned. To prevent the joined ropes from rotating and fatiguing with load changes due to torque mismatch, the p-aramid rope had to be designed with the same torque-tension characteristics as the steel wire rope. The factors determining the torsional response of the rope were studied, the design techniques developed, and the technology experimentally verified.

As a result of this study, a 450-ton break strength rope has now been designed, laboratory tested, and meets all design requirements. A mooring test on an offshore drilling platform is slated to begin shortly.

Overall, mechanical ropes of p-aramid with break strengths from a few kilograms to hundreds of tons are also used in applications such as tennis racket strings, antenna guys, yacht lines, trawl lines, buoy moorings, ship moorings, and two ropes. Electromechanical cables and fiber optic cables reinforced with p-aramid have also been successfully developed for telecommunications systems, both fixed and towed sensor arrays, and as primary umbilicals and tether cables for unmanned undersea work vehicles both for military and commercial applications. In this growing market sector, each use has its own set of requirements, so understanding structure/property relations in order to use the fiber effectively is vitally important.

3.3 Ballistics

3.3.1 Introduction

The properties of p-aramid fibers make them especially suited to stopping ballistic threats and provide a major advance over incumbent ballistic nylon.

Since the U.S. introduction of p-aramid fabric body armor in 1973-1975, over 600 people have been saved from death or serious injury and police fatalities have shown a marked change in trend and numbers. The ability of p-aramid to stop high speed objects is provided by its dynamic energy absorbing properties, primarily its high tensile strength and specific modulus, complemented by good thermal resistance and high fracture toughness.

3.3.2 System Design

Ballistic performance is determined by more than simple fiber toughness or area under the stress-strain curves as shown in Fig. 24. Nylon would have almost twice the ballistic resistance of p-aramid. In practice, p-aramid can absorb about twice the ballistic impact energy of nylon at the same weight, a difference of 4×.

This can be understood by considering the model in Fig. 25 [32]. When a fiber is struck transversely by a high speed projectile, a longitudinal stress wave propagates outward at the speed of sound in the fiber. The speed equals the square root of specific modulus, which for heat treated p-aramid is over 8000 m/sec, 3-4 times that of nylon [33]. It is this wave speed that determines the amount of material that can become involved in the impact event; thus p-aramid is able to involve 3-4 times the volume of material in the

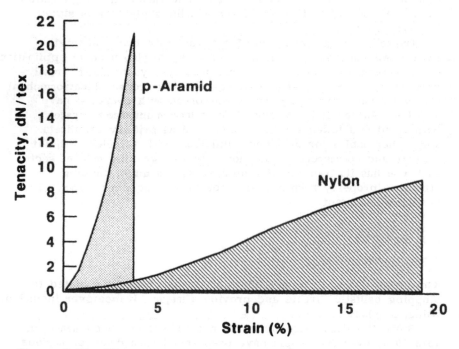

Figure 2.24 Fiber stress-strain curves.

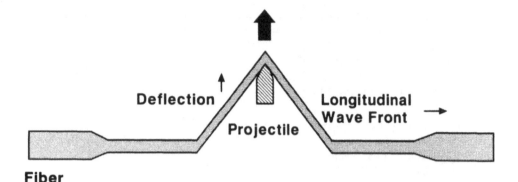

Figure 2.25   Schematic of a fiber being hit by a projectile.

interaction than nylon, resulting in a higher total energy absorbing capacity. The ability to involve a large volume of material because of high specific modulus, therefore, becomes a critical factor.

Also from this model, it is evident that there must be some transverse deflection to load the fiber effectively in tension or the fiber will be sheared and ballistic resistance considerably reduced.

Woven fabrics were found to be a practical product form for ballistics because they give extensive interactions among yarns due to crossovers.

Transverse deflection causes loading of crossover yarns so that up to 50% of the total energy absorption may occur in these secondary yarns. There are negative aspects to crossovers, however, and fabrics must be carefully constructed to balance these effects. For instance, if the weave is too tight or the fabric too stiff, deflection will be restricted, causing shear failure, and stress waves will be reflected back from crossover yarns, intensifying stress at the impact point. Conversely, too loose a weave or too soft a fabric can allow the projectile to penetrate easily by pushing yarns aside or give too much deflection, which can cause serious injuries, called blunt trauma, to the wearer. These and other factors had to be thoroughly understood and technology developed to ensure proper use. This was accomplished working closely with law enforcement agencies and government laboratories.

More recently, p-aramid has been applied in composite form serving both a structural and ballistic function. Design of these systems is more complex than multilayer fabric body armor because of the effects that different resins have on performance and because

Table 2.3 Ballistic Material Comparisons at Equal Target Weight

Fabrics	
Relative energy absorption[a]	
p-Aramid, as-spun	1.0
p-Aramid, heat treated	1.0
Nylon	0.47
Composites (30% polyester)	
p-Aramid, as-spun	0.45
p-Aramid, heat treated	0.40
E-Glass	0.21
Graphite	0.14
Aluminum	0.13

structural requirements often compromise ballistic performance. A detailed discussion of these effects is beyond the scope of this chapter.

Table 3 compares the ballistic resistance of all-fabric and composite armor and p-aramid versus other common armor materials at the same target weight. Here, the threat is metal fragment simulating projectiles in the 300-600 m/sec velocity range, and as-spun p-aramid all-fabric armor is given a relative rating of 1.0.

Today, p-aramid is used as lightweight soft body armor by law enforcement agencies and by civilian and military organizations around the world. The U.S. Army has recently adopted flak vests of p-aramid fiber to replace nylon, giving twice the protection for the same weight. As composite structures, p-aramid is being used in military helmets by the U.S. Army and other armies worldwide. It is also being used as fragment protection for vital components on Navy ships, and for ballistic and nuclear thermal pulse and blast protection on several military shelters in the United States, and is being actively developed for spall liners and various fragment protection uses on military vehicles.

3.4 Asbestos Replacement

3.4.1 Introduction

The story of how p-aramid fiber properties were successfully translated into cost effective, high performance asbestos replacement products is another example of application systems engineering.

In the mid-1970s, manufacturers of friction products—e.g., brakes and clutches—were actively seeking alternatives to asbestos that would be cost effective, processible on existing equipment, perform at least as well as asbestos, and present no health hazards. Many recognized the potential value of the high temperature reinforcing properties of p-aramid, but the cost barrier, 50-100 times the cost of asbestos, seemed insurmountable. Additionally, friction products are composites so totally different from aircraft and marine composites in terms of product forms, processing and fabrication techniques, and performance requirements that a whole new body of technology had to be developed.

One critical step that accelerated the program was the development of a unique engineered form of p-aramid called pulp (Fig. 26). Pulp fibers are short (0.05-8 mm), with many attached fibrils. These fibrils are complex, being curled, branched, and often ribbonlike, and result directly from the inherent fibrillar structure of PPD-T fibers. In composites for friction products the short length provides good dispersibility, and the high surface area (50× standard fiber) and high aspect ratio of the fibrils (greater than 100) provide excellent reinforcement. This, along with thermal stability and toughness, allow p-aramid to replace asbestos in clutches and brakes [34].

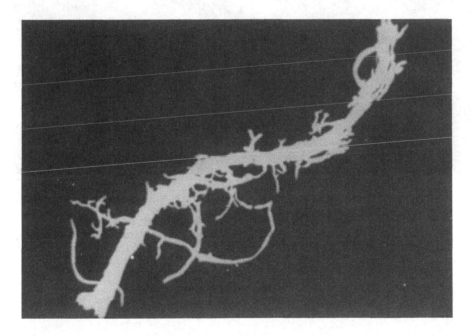

Figure 2.26　Pulp fibers.

(a)

Figure 2.27 Examples of mixer modifications for processing p-aramids: (a) high speed opening blades, (b) optimized plow design, and (c) fluidized bed mixing action.

(b)

(c)

3.4.2 Dispersion Technology

A key technology was discovered with how to open and mix p-aramid pulp with the other components of the friction composite such as resins, fillers, and perhaps other fibers. Uniform distribution of fiber is essential in brakes and clutches to obtain good manufacturing processibility, superior strength, and uniform, predictable wear resistance. The fiber producer, friction product producer, and mixer manufacturers worked together to develop mixer modifications and processing methods acceptable to the friction product's industry (Fig. 27) [35-37].

Demonstration of the value-in-use was also critical to the successful application of p-aramid pulp. Brake formulations, which are complex combinations of a resin, usually phenolic, inorganic fillers, and reinforcing fibers, were researched. By careful selection of inorganic fillers, brakes can be compounded to give substantially improved wear resistance over the asbestos incumbent at p-aramid levels as low as 5-10% of the amount of asbestos needed.

Figure 28 illustrates this improved wear in a high temperature, constant pressure laboratory dynamometer test. These results have been confirmed on vehicle brake tests where, compared to asbestos, p-aramid fiber-reinforced products achieve 2-5 times greater wear resistance in automotive brakes and up to 12 times in truck brakes.

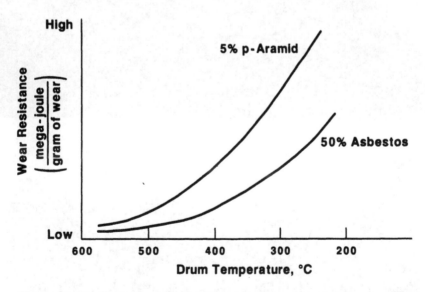

Figure 2.28 Wear resistance of p-aramid and asbestos brake pads at high temperatures.

Thus, p-aramid has proven itself cost effective for friction product uses.

3.4.3 Toxicology

To assure the safety of p-aramid pulp, over 4 years of toxicological testing was necessary. Here also, there were challenging obstacles. Toxicologists had to develop new innovative techniques to generate high levels of respirable sized fiber for meaningful lifetime inhalation studies. Recently, these tests were completed. A 2-year lifetime inhalation study on respirable fibrils with rats confirmed that p-aramid is safe for commercial use and represents no risk to human health [38,39].

Largely as a result of the creative applications technology described here, p-aramid fiber is now used worldwide in automotive, railroad, and industrial friction products as well as in various types of gaskets. The future will see other applications using reinforcing characteristics of pulp products. Reinforcement of sealants and adhesives by pulp is being explored as wet-laid papers. Finally, even more novel, efficient, short fibrous reinforcement forms will be forthcoming.

3.5 Advanced Composite Applications

3.5.1 Introduction

The most significant applications of advanced composites reinforced with p-aramid fibers to date have been in areas where the high strength-to-weight and stiffness-to-weight ratio of the advanced composite materials justifies a cost premium over glass fiber-reinforced composites or conventional materials of construction such as aluminum and steel. Figure 29 illustrates some of the engineering properties of unidirectional epoxy composites of p-aramid, demonstrating that these materials may be the material of choice in many high performance composite applications [40].

One of the primary areas where this is true is in aerospace hardware, specifically in rockets, the Space Shuttle, and in commercial aircraft. These are traditionally areas of very high technology, and to be successful in them, it was necessary to develop new and unique technologies to allow the user to fully capitalize on the unique features of p-aramid. In rockets, filament-wound motor cases reinforced with p-aramid are predominant. The motor case is simply a filament-wound bottle or pressure vessel that contains the solid propellant used to power the rocket. Motor cases must hold the maximum amount of fuel at minimum weight and must sustain the very high pressures generated during the actual burning of the fuel.

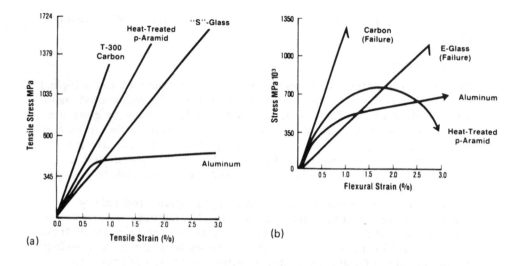

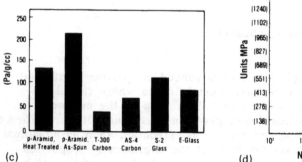

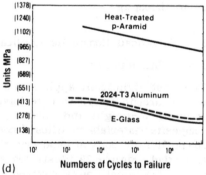

Figure 2.29 Some engineering properties of unidirectional epoxy composites of p-aramid, compared with carbon and glass fiber and aluminum: (a) tensile stress/strain, (b) flexural stress/strain, (c) specific tensile toughness (strand test, and (d) tension-tension fatigue. (From Ref. 40).

3.5.2 Yarn-to-Composite Strength Conversion Efficiency

The figure of merit for motor cases is PV/W, the product of burst pressure times the volume of the bottle divided by its weight. This is analogous to specific strength in other applications. PV/W is affected especially by the strength of the fiber and how well that strength translates in a biaxial stress field (the stresses in the radial and axial direction on the surface of a pressure bottle are of the same order of magnitude).

A comparison of PV/W for materials that were state-of-the-art at the time the present missile systems were under development is shown in Fig. 30 [41]. Note that p-aramid delivers much higher PV/W than either glass, which it replaced, or carbon fiber. With this advantage in PV/W, p-aramid was the logical choice for these systems.

However, it was recognized even then that p-aramid did not perform as efficiently in motor cases as glass or carbon fiber—specifically, the translation of tensile strength from fiber to pressure bottle was lower than for other fibers. To improve the conversion efficiency, the p-aramid fiber was modified specifically for the filament-winding industry.

This new p-aramid fiber gives much better translation of strength into bottles because of a special fiber surface modification that gives the optimum adhesion between fiber and resin in the filament-wound bottle. Because of the high biaxial stresses in a pressure bottle, fibers are stressed not only parallel to but also normal to the fiber axis. By controlling the fiber-to-resin adhesion, it is possible to minimize the stress that the fiber sees normal to its axis, its weak direction.

Figure 31 shows a comparison of the surface-modified aramid with current state-of-the-art high strain carbon fiber, demonstrating superior performance [42].

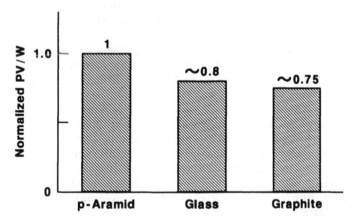

Figure 2.30 Bottle burst comparison for aramid, glass, and carbon fiber; PV/W normalized for aramid = 1.0.

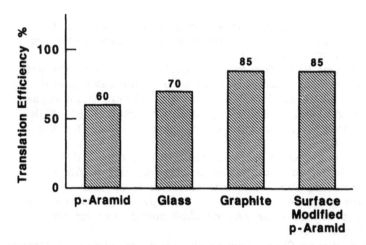

Figure 2.31 Efficiency of strength translation in bottle burst tests of p-aramid, glass, and carbon fibers.

p-Aramid still offers the highest PV/W and, as an added benefit, the already excellent impact damage tolerance of these pressure bottles is further improved by the surface-modified fiber.

3.5.3 Aramid-Carbon Hybrid Composites

In conventional aircraft, impact damage tolerance is also a driving force for the use of p-aramid fiber. In falling-weight impact tests, shown in Fig. 32 [43], the energy required to produce fiber damage has been measured for both p-aramid and carbon fiber reinforcement, showing that p-aramid can absorb roughly twice as much energy without damage.

Large amounts of p-aramid are already being used in commercial aircraft, but almost exclusively in secondary or non-load-bearing structures. In primary load-bearing structures, p-aramid has one major drawback, namely, its compressive strength is much lower than that of either glass or carbon. A technology is being developed to allow aircraft designers to realize both the higher impact damage tolerance and fracture toughness of p-aramid without significant sacrifices in stiffness and compression strength. This technology includes use of hybrids or mixtures of reinforcing fibers.

Hybrids generally offer properties that are intermediate between those of the parent fibers, as is the case with drop-weight impact tests of hybrids similar to those discussed earlier for all-p-aramid and all-carbon fiber. In some cases, however, the two fibers can be combined in such a way as to generate properties superior to those of either parent.

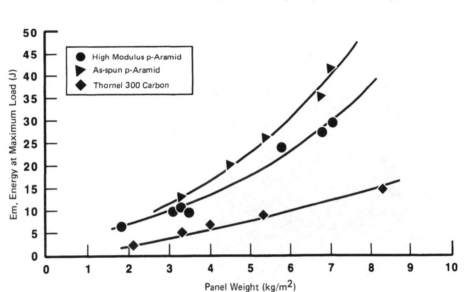

Figure 2.32 Energy required in falling weight tests to produce fiber damage for as-spun and high modulus p-aramid, and carbon fiber-reinforced composite panels.

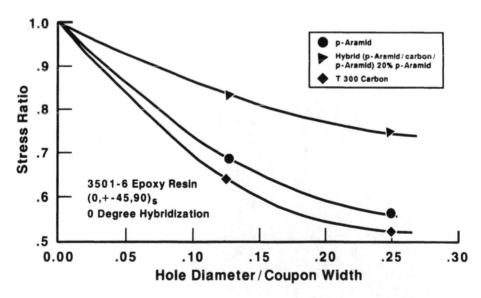

Figure 2.33 Strength of laminates with open holes, with aramid, carbon, and hybrid fabric reinforced.

(a)

(b)

(c)

Figure 2.34 Photographs of crushed tubes (a) reinforced with p-aramid, (b) carbon reinforced composite, and (c) aluminum, showing failure mode.

Figure 33 shows fracture toughness as measured by the reduction in strength of tensile coupons with open holes as a function of the hole diameter for p-aramid, carbon, and a hybrid. Fracture toughness is important in load-bearing structure where there are stress risers such as bolt or rivet holes. The hybrid outperforms the parent fibers significantly because it introduces a new failure mode, namely, interply delamination between plies of carbon and p-aramid, in the vicinity of the hole. This is a very efficient means of redistributing stress around the hole.

In another aircraft application, it is possible to utilize to advantage the compressive properties of p-aramid. Specifically, when crash worthiness and energy absorption during a crash are of concern, the yielding behavior of p-aramid in compression is much preferred to the more brittle, catastrophic failure mode exhibited by carbon fiber. Because it does yield in a manner similar to metals, a composite reinforced with p-aramid has very high energy absorption during crushing and maintains its integrity even after failure.

Figure 2.35 Crushed tube of an aramid/carbon hybrid composite.

This last property is important because the failed structure does not produce fragments that may endanger personnel and because it is still capable of carrying load even after "failure." Thus, it continues to protect the occupants of the aircraft.

In experiments at several laboratories, filament-wound tubes reinforced with p-aramid and carbon fiber were crushed in falling-weight tests, with the results shown in Fig. 34 [44,45]. The highest energy-absorption capability per unit weight of structure is observed for tubes wound with p-aramid near 45° from the tube axis [46].

Where 100% p-aramid has insufficient stiffness, hybrids with carbon are possible that absorb nearly as much energy as all-p-aramid while still enjoying basically the same benign failure mode (Fig. 35).

4. FUTURE DIRECTIONS

p-Aramid technology has expanded rapidly since 1965. This growth will undoubtedly continue on several fronts: improved fiber proper-ties, improved understanding of structure/property relationshps, and new end-use technology—not only for improved performance in existing applications, but also entirely new applications. As some examples in the last category, consider the following items:

1. p-Aramid short fibers and pulp will continue to grow as asbestos replacement materials in brakes, clutches, gaskets, sealants, and adhesives. Advances in these fiber forms will occur to permit increasing value in plastic, cement, and elastomer reinforcement. For example, the tensile properties of natural rubber are changed dramatically by the incorporation of just a few percent of p-aramid pulp. As might be expected from the highly anisotropic nature of the aramid fibers, and the fiber directionality imparted by the rubber milling and extrusion processes, there can be a marked resulting anisotropy of physical properties of the formed elastomer article. Both the potential for improved performance of the end-use article as well as the problems in the fabrication process/article performance optimization are intriguing.

2. As described in Sec. 3.5, the potential exists for synergistic combinations of such dissimilar materials as p-aramid and carbon fiber to generate useful new balances of properties. Extension of the technology will occur with combinations of p-aramid with other fibers. For example, hybrid tire cords with p-aramid and nylon 6,6 yarns have been reported to combine the strength of the aramid fiber with the toughness and elongation of the nylon fiber to generate a whole new range of stress-strain curves, as shown in Fig. 36.

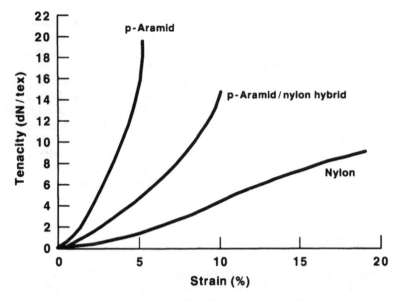

Figure 2.36 Stress-strain curves of p-aramid tire cord, nylon tire cord, and an example of a hybrid combination.

The measured strength of the cord is higher than would be expected from "dilution" of the p-aramid cord by the nylon. There actually is a contribution to the cord breaking strength by the nylon. This phenomenon becomes apparent upon examination of the very smooth stress-strain curves, with no evidence of sequential (i.e., 1, 2, 3) breakage of the individual plys, which is the observed behavior of a nonoptimized cord structure

3.   Use of fiber optic cables for communication is currently small, but growing rapidly. These very delicate glass fiber systems need reinforcement and support to prevent tensile stress on the glass filaments, especially during installation and sefvicing. The high modulus, low stretch form of p-aramid fiber is the material of choice in this application due to its light weight, high resistance to stress, and excellent tensile fatigue performance.

Figure 37 shows how the p-aramid fiber is incorporated into the fiber optics system.

4.   As a final example of the demand that evolving technologies are placing on the need for reinforcement materials with unique properties, consider a trend in the computer industry. The need for faster and faster processing of data is growing to accommodate the explosion in information processing. In many cases, the speed of data processing is limited not only by the capability of the micro-

chips used by the computer, but now even by the proximity of the components in the computer—the distance electric impulses have to travel on the circuit board. The advent of a new generation of microcircuits encapsulated in ceramics requires a circuit board with a matching coefficient of thermal expansion in order to pack the chips as close together as possible, and to use plug-in sockets.

Aramid fibers have a very low coefficient of thermal expansion, negative along the fiber axis, and positive across the fiber diameter. This is undoubtedly due to the extended chain structure and high orientation of the PPD-T molecule, and hence the difference in interatomic forces along the C axis of the unit cell (molecular axis) and the A and B axes (hydrogen bonding and coordinate bonding axes); see Fig. 3. When the fibers are imbedded in an epoxy material, for example as in a fabric, the fiber coefficient of expansion balances the positive coefficient of the matrix to give a composite board with a net coefficient of expansion of essentially zero in the X-Y plane of the composite. Since the chips also have essentially a zero coefficient of expansion, there is an almost perfect match,

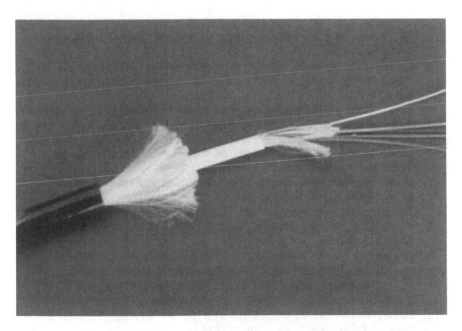

Figure 2.37 Fiber optic cable, with the layers cut away to show the components. The high modulus aramid reinforcement is the outer layer of filaments.

permitting very high density chip packing on the circuit board, without danger of stressing or breaking the now very short, delicate wires connecting the chip to the board circuitry.

There are many more new and exciting applications that are developing. The potential for these materials to generate new systems with new performance levels is limited only by the designer's imagination.

ACKNOWLEDGMENT

Much of the material in this chapter was the subject of a presentation at the International Symposium on Fiber Science and Technology, August 20-24, 1985, Hakone, Japan by D. Tanner, J. A. Fitzgerald, W. F. Knoff, and J. J. Pigliacampi, "Aramid Fiber Structure/Property Relationships and Their Applications to Industrial Materials."

REFERENCES

1. M. Jaffe, and R. S. Jones, in *High Technology Fibers, Part A* (M. Lewis and J. Preston, Eds.), Marcel Dekker, New York, 1985, chap. 9.
2. J. Preston, Aramid Fibers, in *Kirk-Othmer Encyclopedia of Chemical Technology*, 3rd ed., vol. 4, Wiley-Interscience, New York, 1978.
3. P. J. Flory, Molecular Theory of Liquid Crystals, in *Advances in Polymer Science*, Vol. 59, Springer-Verlag, Berlin, 1984, pp. 1-36.
4. S. P. Popkov, Liquid Crystalline Order in Solutions of Rigid Chain Polymers, in *Advances in Polymer Science*, vol. 59, Springer-Verlag, Berlin, 1984, pp. 76-99.
5. P. W. Morgan, *Macromolecules*, 10, 1381 (1977).
6. S. L. Kwolek, P. W. Morgan, J. R. Schaefgen, and L. W. Gulrich, *Macromolecules*, 10, 1390 (1977).
7. T. I. Bair, P. W. Morgan, and F. L. Killian, *Macromolecuels*, 10, 1396 (1977).
8. H. Blades, U.S. Patent 3,767,756 (to du Pont), 1973.
9. H. Blades, U.S. Patent 3,869,429 (to du Pont), 1975.
10. M. G. Northolt, *Eur. Polym. J.*, 10, 799 (1974).
11. J. W. Ballou, *Polym. Prep.*, 17, 75 (1976).
12. M. G. Dobb, D. J. Johnson, and B. P. Saville, *Polym. Sci., Polym. Phys. Ed.*, 15, 2201 (1977).
13. R. Hagege, M. Jarrin, and M. Sotton, *J. Microsc.*, 115, 65 (1979).

14. G. S. Fielding-Russell, *Text. Res. J.*, *41*, 861 (1971).
15. K. E. Perepelkin and Z. U. Chereiskii, *Mekhanika Polimerov*, *6*, 1002 (1977).
16. T. Ito, *Sen i Gakkaishi*, *38*, 54 (1982).
17. M. G. Northolt and R. v. d. Hout, *Polymer*, *26*, 310 (1985).
18. W. B. Black and J. Preston, *High Modulus Wholly Aromatic Fibers*, Marcel Dekker, New York, 1973.
19. I. M. Brown, T. C. Sandreczki, and R. J. Morgan, *Polymer*, *25*, 759 (1985).
20. M. G. Dobb, D. J. Johnson, and B. P. Saville, *Polymer*, *22*, 960 (1981).
21. H. Blades, U.S. Patent 3,869,430 (to du Pont), 1975.
22. F. J. Kovac, in *Tire Technology*, 5th ed., The Goodyear Tire and Rubber Company, Okron, Ohio, 1978, chap. 3.
23. J. Zimmerman, *Text. Manuf.*, *101*, 49 (1974).
24. R. E. Wilfong and J. Zimmerman, *J. Appl. Polym. Sci.*, *Appl. Polym. Symp.*, *31*, 1 (1977).
25. M. M. Horn, Strength and Durability Characteristics of Ropes and Cables from Kevlar® Aramid Fibers, Oceans '77 Conference Record, Conference sponsored by Marine Technology Society and IEEE, October 17-19, 1977, Los Angeles, Calif.
26. T. J. Kozik and J. Noerager, Riser Tensioner Force Variations, Paper OTC 2648, Offshore Technology Conference, Houston, Tex., 1976.
27. M. A. Childres and E. B. Martin, Field Operation of Drilling Marine Risers, *Proc. 1978 European Offshore Petroleum Conference*, London, October 24-27, 1978.
28. P. T. Gibson, Analytical and Experimental Investigation of Aircraft Arresting Gear Purchase Cable, NTIS Rept. AD 904263, Battelle Memorial Institute, Columbus. Ohio, July 3, 1967.
29. P. T. Gibson, Continuation of Analytical and Experimental Investigation of Aircraft Arresting Gear Purchase Cable, NTIS Rept. AD 869092, Battelle Memorial Institute, Columbus, Ohio, April 8, 1969.
30. M. A. Childers, Deep Water Mooring—Environmental Factors Control Station Keeping Methods, *Petroleum Eng.*, September 1974.
31. M. A. Childers, Deep Water Mooring—The Ultradeep Water Spread Mooring System, *Petroleum Eng.*, October 1974.
32. D. Roylance and S. S. Wang, Penetration Mechanics of Textile Structures, NTIS Rept. AD-A089445, U.S. Army Natick R&D Dommand, Natick, Mass., June 1979.
33. C. E. Morrison, and W. H. Bower, Factors Affecting the Ballistic Impact Resistance of Kevlar® Laminates, in *Advances in Composite Materials—Proceedings of the Third International Conference*

of Composite Materials, Paris, 26-29 August 1980, Pergamon Press, Oxford, 1980.

34. H. Y. Loken, Asbestos Free Brakes and Dry Clutches Reinforced with Kevlar® Aramid Fiber, Society of Automotive Engineers paper 800667, Earthmoving Industry Conference, Peoria, Ill., April 14-16, 1980.

35. Littleford Processing of Kevlar® Aramid Fiber, Bull. LM-226, Littleford Bros., Florence, Ken., 1982.

36. Eirich Processing Equipment for Kevlar® Aramid Fibre and Pulp, Bull. EM-505, Eirich Machines, New York, 1983.

37. Guide to Processing Kevlar® Aramid Fiber and Pulp for Friction Products, Bull. E-65333, E. I. du Pont de Nemours & Co., Wilmington, Del., June 1984.

38. K. P. Lee, D. P. Kelly, and G. L. Kennedy, Jr., *Toxicol. Appl. Pharmacol.*, *71*, 242 (1983).

39. Environmental Protection Agency, Log No. 8EHQ-0485-0550, April 4, 1985.

40. D. Tanner, J. L. Cooper, A. Dhingra, and J. J. Pigliacampi, Future of Aramid Fiber Composites as a General Engineering Material, *J. Jap. Soc. Comp. Matl.*, *11*, 196 (1985).

41. O. C. Wright, AIAA Paper 73-1259, AIAA/SAE 9th Propulsion Conf., Las Vegas, Nev., 1973.

42. G. E. Zahr, An Improved Aramid Fiber for Aerospace Applications, in *Progress in Advanced Materials and Processes: Durability, Reliability and Quality Control*, Elsevier, Amsterdam, 1985.

43. M. W. Wardle and E. W. Tokarsky, Drop Weight Testing of Laminates Reinforced with Kevlar® Aramid Fibers, E-Glass and Graphite, *Comp. Technol. Rev.*, ASTM, Philadelphia, Spring 1983.

44. M. W. Wardle, Designing Composite Structures for Toughness, Design and Use of Kevlar® Aramid in Composite Structures, Technical Symposium V sponsored by E. I. du Pont de Nemours & Co., Reno, Nev., June 1984.

45. J. D. Cronkite, "Design of Helicopter Composite Structures for Crashworthiness, Design and Use of Kevlar® Aramid in Composite Structures, Technical Symposium V sponsored by E. I. du Pont de Nemours & Co., Reno, Nev., June 1984.

46. K. Sen Joyanto, Designing for a Crashworthy All-Composite Helicopter Fuselage, *Proc. 40th Annual Forum American Helicopter Society*, Arlington, Va., May 16-18, 1984.

3

FIBERS FROM NAPHTHALENE-BASED THERMOTROPIC LIQUID CRYSTALLINE COPOLYESTERS

MICHAEL JAFFE, GORDON CALUNDANN and HYUN-NAM YOON /
Hoechst Celanese Corporation, Summit, New Jersey

1. INTRODUCTION

Since the introduction of the high modulus aramid fiber, Kevlar, by du Pont in the late 1960s [1], many research groups worldwide have searched for improved or more cost-effective routes to similar or superior fiber properties. It had been recognized since the pioneering work of Herman Mark that the modulus of a polymer in the solid state is a function of three parameters:

The inherent modulus of the molecular chain
The chain packing density
The perfection of molecular orientation

By about 1970, it was recognized that polymers which demonstrated nematogenic properties greatly simplified the achievement of the high degree of molecular order necessary in the solid state to approach theoretical tensile property levels. This is illustrated in Fig. 1, which contrasts the fiber forming process for conventional and nematic polymers in terms of microstructure. The aramids are lyotropic, exhibiting nematic behavior in solution (100% sulfuric acid is the solvent of choice) [1-3]. The potential for process simplification and flexibility possible through the substitution of melt for solution spinning—i.e., the utilization of a thermotropic nematogen for the lyotropic aramids—is self-evident. The announcement of

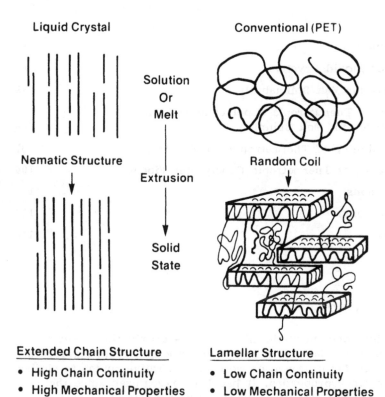

Figure 3.1 Structure development during fiber forming process for conventional and nematic polymers.

the thermotropic polyester system, copoly-(ethylene terephthalate/p-oxybenzoate), by the Eastman Company in 1976 [4-6] demonstrated the feasibility of this approach. Over the past decade a variety of thermotropic copolyesters have been identified, patented, and evaluated as fibers. For reviews of this technology see, for example, the work of Calundann [7,8], Jaffe [9,10], and Schaefgen [11]. As of mid-1987, none of these fibers was truly commercially available, although several were in various stages of commercial development. The purpose of this chapter is to review the underlying science and technology of fibers derived from thermotropic copolyesters, with emphasis on current status and criteria for commercial success.

2. POLYMER LIQUID CRYSTALS

The three basic liquid crystalline textures, all of which have been observed with polymeric molecules, are illustrated in Fig. 2. Of these, only the nematic texture offers the highly oriented, non-chain-end registered structure in the solid state desirable for high modulus fiber applications. The nematic state in polymers can be achieved in three ways: rigid, planar moieties in the mainchain backbone; rigid units in the backbone separated by "flexible spacer" units; and nematogenic side chains off a flexible mainchain. These three approaches are shown diagramatically in Fig. 3. Considerations of chain stiffness and packing density lead to the conclusion that only the rigid main chain architecture can lead to ultra-high mechanical property levels. A characteristic of nematic melts is the phenomena of melt anisotropy, the persistence of orientational order in the melt in the absence of external fields. Figure 4 illustrates the birefringent behavior of a typical polymer nematic melt between crossed polars. The black "threads" are the orientational domain boundaries always associated with nematic melts.

A further highly desirable feature of polymeric nematic liquid crystals as fiber formers is the ease with which they orient in elongational flow fields, exhibiting low viscosities even at high molecular weights. The origin of this behavior lies in the "pencils in a box" structure of the nematic melt and the ease with which the "pencils" slide by one another when desormed along their long axis. Figure 5 illustrates the shear rate dependence of the viscosity of a typical thermotropic copolyester. In contrast to conventional polymer melts, the shear viscosity of thermotropic copolyesters is dependent on shear rate over many orders of magnitude. Even at very low shear rates, no zero shear rate viscosity region is observed, resulting in very low viscosities at typical fiber formation conditions. The rheology of liquid crystal polymers has been extensively reviewed by Wissbrun [12].

Nematic

- Molecules Show Parallel One-Dimensional Order
- Turbid Liquid
- Low Viscosity

Smectic

- Molecules Align Parallel And Stratified; Two-Dimensional Order
- Turbid Liquid
- Highly Viscous

Cholesteric

- Shown By Optically Active Molecules Only
- Nematic Layers Arranged In Helical Structure
- Iridescent Liquid With Optical Rotatory Power
- Highly Viscous

Figure 3.2 Texture and physical characteristics of mesomorphic phases.

- **Rod-Like**

  Aromatic Polyamides, Esters, Azomethines, Benzbisoxazoles

- **Helical**

  Polypeptides, Nucleotides, Cellulosics

- **Side Chain Mesogenic ("Comb" Polymers)**

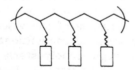

- **Block Copolymers (Alternating Rigid/Flexible Units)**

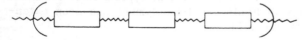

Figure 3.3 Molecular architecture for liquid crystal polymers.

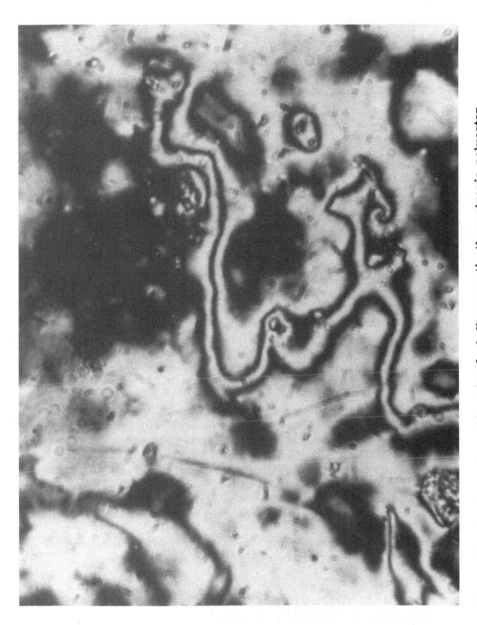

Figure 3.4 Texture of nematic melt of wholly aromatic thermotropic polyester.

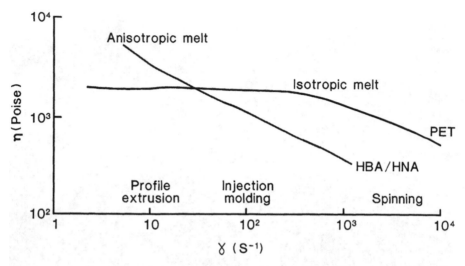

Figure 3.5   Rheology of the nematic melt of a HBA/HNA copolyester.

3.   CRITERIA FOR MELT SPINNABILITY

For a thermotropic copolyester composition to be a viable candidate as a melt spinnable, commercial high performance fiber precursor, several criteria must be met. These include:

Melt stability—a window must exist between the melting and decomposition temperatures associated with the backbone molecular structure to allow for stable processing.

Melt anisotropy—the backbone must be designed to provide the desired rheology in the temperature range of interest, ideally between 250 and 300°C.

Molecular design—the backbone architecture must pack in the solid state in a manner that yields the desired solid state properties.

Monomer cost—the monomers must be cost effective when compared to competitive raw materials and processes.

4.   DESIGN OF THERMOTROPIC COPOLYESTER MOLECULAR STRUCTURES FOR MELT SPINNING

By 1974 the known wholly aromatic polyesters fell into one of two groups. The first category consisted of the highly crystalline, very high melting, and essentially intractable materials such as poly(p-oxybenzoate) or poly(p-phenylene terephthalate). These polymers are not melt processable, although compression molding and plasma spray fabrication techniques were and are used with some success. An exception within this category was the copolyester of p-hydroxybenzoic acid, terephthalic acid, and 4,4'-biphenol, a Carborundum polymer with the trade name Ekkcel I-2000. This polymer is injection moldable at temperatures in the vicinity of 400°C, a temperature not compatible with common melt spinning [4] equipment. Polyester thermal degradation at 400°C makes stable fiber production particularly difficult. The second category of polyarylates well known at the time included the amorphous or nearly amorphous polyesters. These polymers formed clear glassy solids from isotropic melts and are generally synthesized with monomer formulations containing relatively asymmetric molecules such as isophthalic acid, resorcinol, or bisphenol-A. The glassy polyarylates are extremely difficult to orient and have, not surprisingly, unexceptional tensile properties.

A 1972 patent to Cottis et al. [4] compared melting point-composition curves of two wholly aromatic copolyesters: the copolymer of p-hydroxybenzoic acid (HBA), terephthalic acid (TA), and hydroquinone (HQ), and the copolymer made from HBA, TA, and 4,4'-biphenol (BP), that is, Carborundum's Ekkcel I-2000 polymer (Fig. 6).

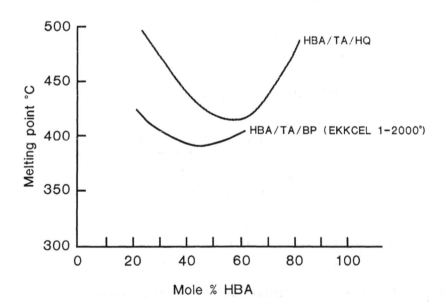

Figure 3.6 Melting point-composition curves of two wholly aromatic copolyesters: HBA/TA/HQ and HBA/TA/BP.

The minimum melting temperature of the former copolymer is about 420°C, while the biphenol based, injection moldable copolyester has a minimum melting temperature of about 395°C. In 1976, in another patent to Carborundum [5], an HBA/TA/BP copolymer system, softened with isophthalic acid (IA), was reported to be melt spinnable to produce fiber with an as-spun tensile modulus of 390 g/d (ca. 7 Msi). An annealing and draw process improved these properties to a reported 780 g/d (ca. 14 Msi). While these patents did not mention thermotropic behavior, later work has shown that some compositions of this terpolyester form nematic melts.

Portions of this patent property were subsequently sold to Dart-Kraft Corporation to become the basic technology supporting the engineering resin, Xydar, of the newly named Premark International Corporation. The technology was recently acquired by Amoco.

Figure 7 summarizes some key points on what was the first reported and well-characterized thermotropic polymer system. These copolyesters, made from p-acetoxybenzoic acid and PET, form a series of thermotropic aliphatic-aromatic polyesters, the so-called

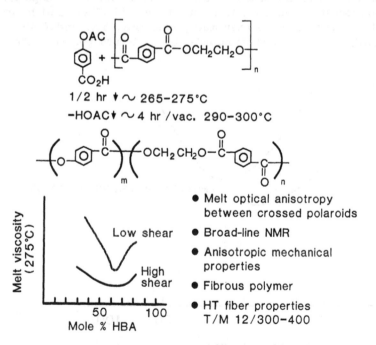

Figure 3.7   Summary of preparation method and physical characteristics of a thermotropic copolyester system based on p-acetoxybenzoic acid and poly-(ethylene terephthalate).

X7G polymers [6]. Many of these compositions showed the melt anisotropy associated with nematogenic polymers. Broad-line NMR [13], relating melt molecular order with polymer composition and temperature, also provided evidence supporting the thermotropic nature of these materials. Facile and high polymer orientation is noted for these polymers in the direction of flow, and polymer melts between crossed polars show typical nematic texture. Reduced melt viscosities were found for polymers containing from 40 to perhaps 70 mol% p-oxybenzoyl. The lower melt viscosities, with a minimum at about 60-70 mol% HBA content, are additional evidence supporting a nematic melt structure for the copolyesters within these compositional limits. The fiber properties reported for Eastman's oxybenzoyl-PET copolyesters are not particularly impressive, with tensile moduli in the range of 300-400 g/d and maximum tensile strengths rarely exceeding 12 g/d.

Early in 1986, Unitika Limited-Nippon Telephone and Telegraph [14] and Mitsubishi Chemical Industries [15] announced the intention of commercializing fibers based on this technology. Eastman's longer range intentions with X7G-type polymers as engineering resins are unclear, although it is known that some internal and market development activities are continuing.

Figure 8 summarizes the molecular architectures used to promote polyester melt anisotropy at reasonable temperatures. The Celanese, du Pont, and Eastman laboratories have outlined most of the key approaches to lower polymer melting point via crystalline order disruption, although research remains ongoing at these and other industrial organizations [7,8].

Polymer tractability is accomplished while maintaining sufficient molecular symmetry to preserve the melt anisotropy inherent in linear aromatic polyesters, that is, polyesters derived from symmetric monomers such as p-HBA, TA, HQ, BP, and the like, as follows. The Eastman tack, already reviewed [6,13], involved the introduction of aromatic monomers into aliphatic-aromatic polyesters such as PET. The increased symmetric ring content imparts melt anisotropy to the copolyester, but the aliphatic-aromatic structure does not yield the mechanical property levels obtained with the wholly aromatic polyesters.

Introduction of bent rigid moieties, such as from isophthalic acid or resorcinol, is an obvious route toward improving polymer tractability. There are problems, however, associated with the meta linkage. For example, incorporation of low levels of isophthalic acid provides an increase in polymer tractability, but increasing isophthaloyl content beyond a certain point tends to offset this gain by reducing polymer melt anisotropy with consequent negative impact on polymer rheology and fiber properties. Resorcinol incor-

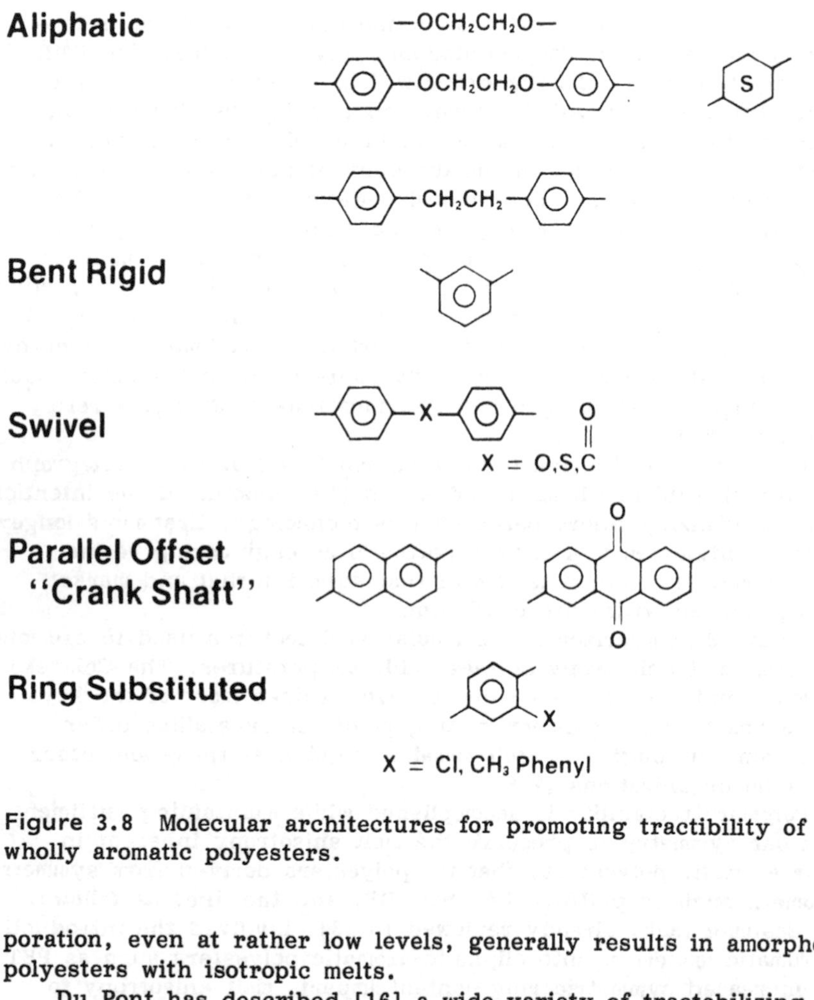

Aliphatic

$$-OCH_2CH_2O-$$

Bent Rigid

Swivel

X = O,S,C

Parallel Offset "Crank Shaft"

Ring Substituted

X = Cl, CH₃ Phenyl

Figure 3.8 Molecular architectures for promoting tractibility of wholly aromatic polyesters.

poration, even at rather low levels, generally results in amorphous polyesters with isotropic melts.

Du Pont has described [16] a wide variety of tractabilizing monomers and the thermotropic polyesters derived therefrom. Research has focused on ring substituted monomers such as chloro, methyl, or phenyl substituted hydroquinone and "swivel" or linked ring molecules, examples of which are 3,4' or 4,4' functionally disubstituted diphenyl ether, sulfide, or ketone monomers. Use of the parallel offset or "crankshaft" geometry provided by 2,6 functionally disubstituted naphthalene monomers has been the major thrust of the Hoechst Celanese development of wholly aromatic thermotropic polyesters [17,18].

Specifically, Hoechst-Celanese Corporation has defined families of thermotropic copolyesters based on 2,6-naphthalene dicarboxylic acid (NDA), 2,6-dihydroxynaphthalene (DHN), and 6-hydroxy-2-naphthoic acid (HNA). Based on this chemistry, Hoechst Celanese has recently (1985) commercialized a family of liquid crystalline engineering resins under the trade name Vectra, and in 1986 announced a joint feasibility study with the Kuraray Company to commercially evaluate such materials as high performance fibers.

To illustrate the dramatic effect on polymer melting point depression of the 2,6-naphthalene nucleus, Fig. 9 compares the T_m-copolyester composition curves of NDA, DHN, and HNA based polymers with those of the copolyester HBA/TA/HQ. Thus, replacement of TA with the 2,6-naphthalene diacid gives a series of copolyesters with a melting point minimum at about 325°C, near 60 mol% HBA; the 2,6-naphthalene diol, replacing hydroquinone, produces a series with a polymer T_m low near 280°C, and much of the composition range of the two-component polyester of HBA and HNA falls within the industrially convenient melt temperature zone of 250-310°C. A great many compositions of the three groups of copolyesters shown were investigated in detail; all formed anisotropic melts, and all were melt spun to high strength, high modulus fibers.

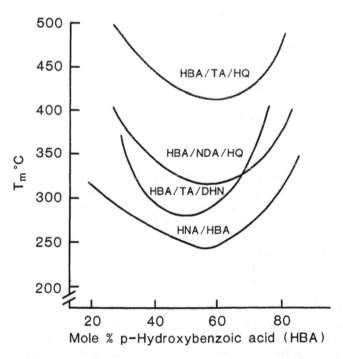

Figure 3.9 Melting point-composition curves of copolyesters containing 2,6-naphthylene moieties.

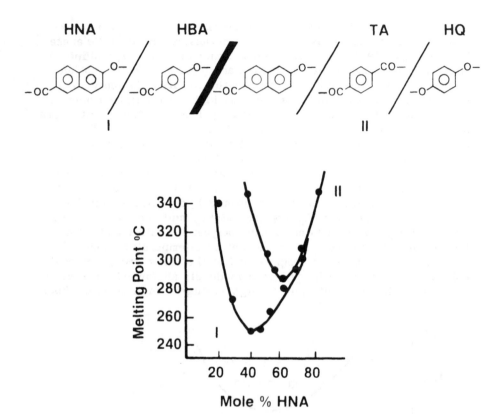

Figure 3.10 Melting point-composition curves for copolyester systems based on HBA/HNA and HNA/TA/HQ.

Replacement of the p-oxybenzoyl fraction with p-phenylene terephthaloyl units in the two-component copolyester results in a more symmetric, higher melting copolyester series, shown as polymer II, HNA/TA/HQ, in Fig. 10. Similarly, all of the compositions of II that were examined showed nematic melts and gave high strength, high modulus fibers.

As mentioned, the influence of the meta linkage on a well-defined thermotropic polyester system is mixed. When isophthalic acid was interpolymerized with HBA, NDA, and HQ, a four-component terpolyester, HBA/NDA/IA/HQ [18], with increased tractability over the HBA/TA/HQ copolymer, was obtained. Figure 11, the ternary T_m-composition diagram of this terpolymer, shows contour plots representing groups of terpolyester compositions with the same melting point. The shaded area encloses those terpolyesters of close to the

optimum processability-fiber property profile, that is, systems melting at 300°C or less, with equal to or less than 15 mol% isophthaloyl
units and from 55–75 mol% p-oxybenzoyl units. Above 30°C, stable
fiber spinning becomes increasingly difficult. Note, however, that
lower melting terpolyesters of increased isophthaloy content show
less melt anisotropy and thus less favorable rheology. Fiber from
higher isophthaloyl terpolyesters has a reduced tensile property
range combined with lower chemical stability.

Work with NDA, DHN, and HNA monomers has also led to the
development of thermotropic aromatic poly(ester-amides) at Hoechst-
Celanese. It was reasoned that the introduction of hydrogen bonding

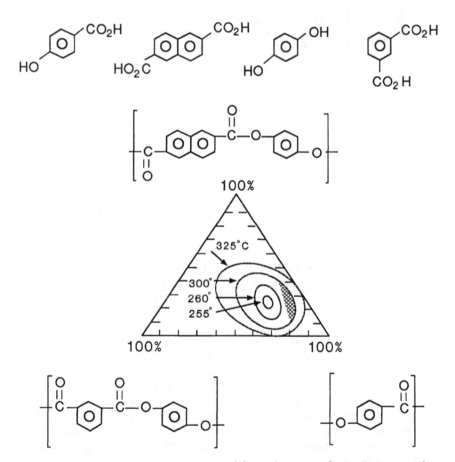

Figure 3.11 Melting point-composition diagram of wholly aromatic
copolyesters containing HBA, HNA, HQ, and IA.

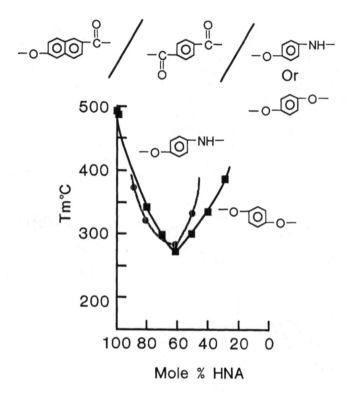

Figure 3.12   Comparison of melting point-composition behavior of
HNA/TA/HQ and analogous poly-(ester amide).

into the chain might result in useful property improvements, i.e.,
increased fiber fatigue resistance, higher filament shear modulus
(and tensile modulus), and improved fiber surface adhesion. Early
data are promising, and ongoing research seeks to define the potential of these ester-amide variants. However, the introduction of
a mixed linkage does not appear to offer any advantage in polymer
tractability. Figure 12 compares the T_m-composition curves of the
HNA/TA/HQ copolyesters with those of the analogous poly(ester-
amide)s, in which the p-acetoxyacetanilide replaces hydroquinone
diacetate. In fact, amide linkage incorporation appears to reduce
the tractable range, although minimum melting points for both systems
were nearly the same at about 275-280°C (about 60 mol% HBA). The
maximum amide linkage tolerable from a tractability standpoint was
about 25 mol% [19].

   While the dominant patent positions in high performance, thermotropic polyester chemistry were staked out by Hoechst-Celanese,

Carborundum, du Pont, and Eastman, research and consequent patent and test market activities on derivative polymer compositions continues worldwide. In fact, several of these newer polymers are now in the commercial development stage, although most are directed at engineering resin rather than fiber applications. An exception to this trend is Carborundum's IA modified, 4,4'-biphenyl based wholly aromatic polyester. This polymer, now owned by Dartco Manufacturing (Premark International, Inc.), was licensed in 1987 to Allied Fibers for development as a high strength, migh modulus fiber. In Japan, Sumitomo had also announced intentions to develop this polymer as a fiber but altered its strategy to emphasize the composition in engineering resin markets. Unitika and Mitsubishi are assessing the potential of X7G-like compositions as fibers for the Japanese market. Unitika is also developing a wholly aromatic thermotropic polyester of propriatory composition, while MCI concurrently pursues a liquid crystalline poly(ester-imide) for resin and possibly fiber applications.

ICI (U.K.), after researching the technology for many years, has commercialized its Victrex SRP1 and SRP2 thermotropic polyesters as engineering resins. One of these polymers is believed to be based on HBA/IA/HQ polymers, modified and made more processable by the low level incorporation of 6,2-hydroxynaphthoic acid (HNA).

In addition to Hoechst AG, via the 1987 acquisition of Celanese Corporation, the other West German chemical giants have recently entered the liquid crystal polymer (LCP) arena with internally developed candidates. BASF, working with compositions based on biphenol, has introduced the Ultrax family of engineering resins. Bayer is sampling its KV series of LCP resins. These materials appear to be thermotropic liquid crystalline polyester carbonates. It is not clear at this time whether BASF or Bayer plans to commercialize their respective LCPs in fiber form.

5. MELT SPINNING OF THERMOTROPIC COPOLYESTERS

A typical fiber spinning process encompasses two distinct melt flow fields; a shear deformation (capillary flow in the spinning jet) and an elongational deformation (melt drawdown). Available data suggest that the high orientation in LCP fibers results from the elongational flow rather than the shear flow in the spinning process. This can be seen clearly in Fig. 13 [20], which shows the variation of fiber radius along the spinline axis in the melt drawdown region, along with the wide-angle X-ray diffraction pattern at selected points along the spinline. The data show that LCP melts exhibit a moderate

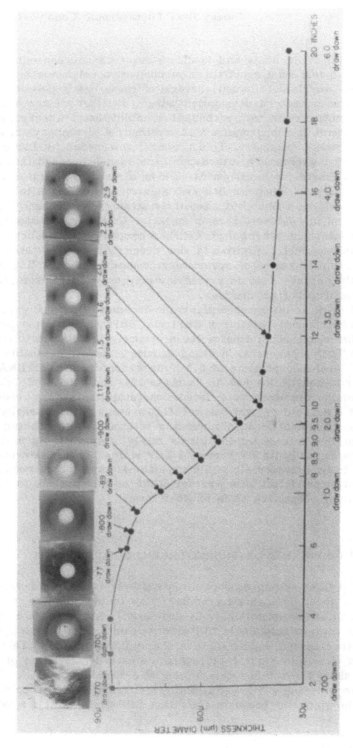

Figure 3.13 Orientation development in an HBA/HNA copolyester fiber during fiber spinning process.

extent of die swell and that the drawdown process is confined to a narrow region. Furthermore, the diffraction data clearly demonstrate that the melt exiting from the die does not develop appreciable chain orientation (despite a considerable amount of shear deformation— a typical shear rate in the spinning die is of the order of 10^4 sec^{-1}). In contrast, the orientation development in the elongational flow field associated with melt drawdown is extremely efficient; at a drawdown ratio of 3.0, the diffraction pattern of the fiber is almost indistinguishable from that of a fully drawn fiber.

Rapid axial chain orientation development with melt drawdown manifests in the rapid development of fiber modulus in the spinline. As the data in Fig. 14 demonstrate, the LCP fiber modulus reaches an asymptotic value at a drawdown ratio of about 10, and further drawdown of the fiber does not improve the modulus. This indicates that the fiber orientation is fully developed at this drawdown ratio, although further reduction in the fiber diameter takes place. Furthermore, the fiber modulus cannot be further improved with subsequent processing steps such as cold-drawing or annealing.

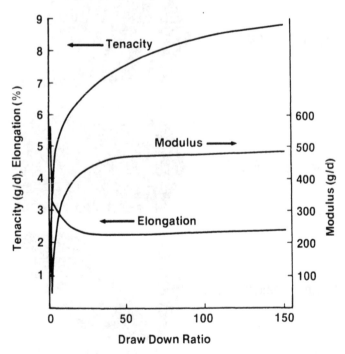

Figure 3.14 Fiber property development of a HBA/HNA fiber as a function of melt drawdown.

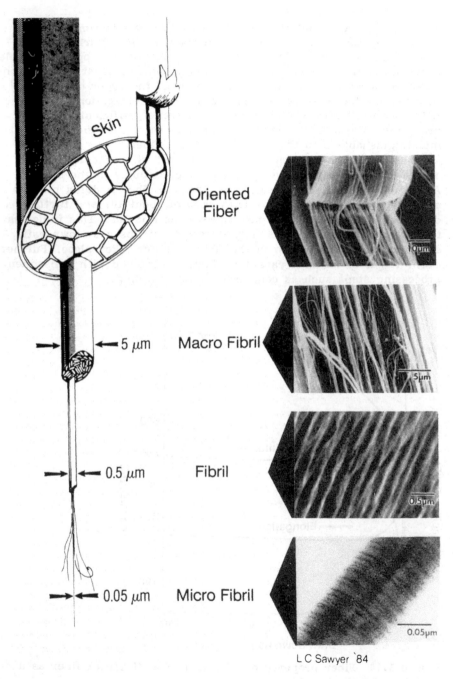

Skin

Oriented
Fiber

←5 μm Macro Fibril

←0.5 μm Fibril

←0.05 μm Micro Fibril

L C Sawyer `84

Figure 3.15 Microscopic structure of a highly oriented thermotropic fiber.

It has to be stated that the excellent mechanical properties
of wholly aromatic, thermotropic copolyester fibers arise primarily
from the easy orientability of the mesophase and the resulting high
molecular chain orientation. Structure analysis of the fibers using
thermal analysis, wide-angle X-ray diffraction, and electron microscopy reveals that the molecular organization in as-prepared fibers
is, to a first approximation, that of a polymeric glass with a nematic
structural order and high degree of axial chain orientation, i.e.,
an array of well-packed rods. The solid state microstructure is
highly fibrillar and is best described as a hierarchy of distinct
fibrillar species, as shown in Fig. 15 [21,22]. This microstructure
is valid for all thermotropic copolyesters investigated and appears
quite a general description of highly oriented polymers in the solid
state [21].

X-ray fiber diagrams of a series of copolyester fibers with
compositions of 30/70, 58/42, and 75/25 HBA/HNA are shown in
Fig. 16. Note that the positions of all observed intensity maxima
are shown in Fig. 14 [20]. The fiber diagrams show intense broad
equatorial scatter, together with well-defined meridional maxima.
Such diagrams are characteristic of highly oriented, parallel arrays
of polymer chains that have poor lateral registry of atoms normal
to the chain direction, probably due to axial shifts and stacking
faults. Although these data suggest that there may be some three-
dimensional order in the fiber structure, the diffuse equatorial
scatter implies that much of the lateral packing of the chains in
these fibers is irregular. With a few exceptions, all fibers of wholly
aromatic copolyesters exhibit a similar fiber diagram [21,24-26].

The nature of the chain packing in thermotropic copolyester
fibers has been investigated by Blackwell et al. [24,26,31]. Comparing

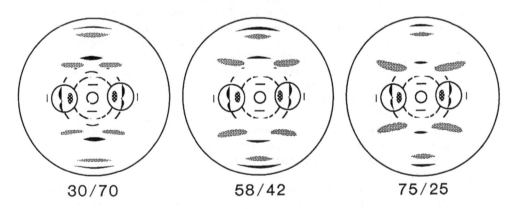

30/70   58/42   75/25

Figure 3.16 Wide-angle X-ray diffraction diagrams of fibers from
30/70, 58/42, and 75/25 HBA/HNA copolyesters.

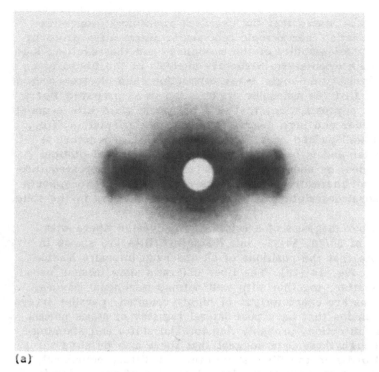

(a)

Figure 3.17 Effects of annealing on wide-angle X-ray diffraction diagram of fibers of 25/75 HBA/HNA copolyester: (a) as-spun, (b) annealed.

the observed meridional intensity maxima with theoretical models for the diffraction patterns expected for random copolymers, it was concluded that the meridional maxima seen with thermotropic copolyester fibers represent the distribution of monomer sequences along the polymer chain. Typical diffraction patterns are shown in Fig. 13. The merional maxia are aperiodic; i.e., they are not orders of a simple polymer repeat, and they change in number and position with monomer type and ratio.

These results confirm that the thermotropic copolyesters are composed of chains of random monomer sequence and with little positional correlation between the atoms in neighboring chains. It may be inferred that the interchain interaction in the fibers is relatively weak.

The perfection of the packed-rod nematic structure of the fibers improves with annealing close to the melting point. The X-ray diagrams of as-prepared and annealed fibers of HBA/HNA copolyesters

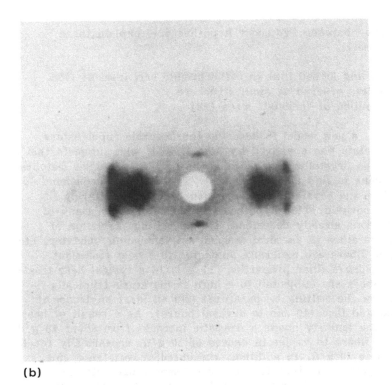

(b)

are shown in Fig. 17 [27]. Significant diffraction pattern changes
are the following:

> The broad equatorial scatter separates into two well-defined
> maxima.
> The residual scattering around the beamstop decreases.
> The meridional maxima sharpen in both directions.

Most of the thermotropic copolyester fibers do not show any
discernible increase in chain orientation with annealing. The observed
changes in the fiber diagram are interpreted as the chains assuming
a well-defined orthorhombic arrangement with annealing from the
pseudo-hexagonal oriented nematic structure associated with as-spun
fibers.

Annealing also tends to increase the DSC melting point and
sharpen the melting endotherm. Available data are not sufficient

to discriminate between two major hypotheses to explain these
changes, namely:

Chains being locked into an orthorhombic arrangement from
the less ordered as-spun structure
Redistribution of "crystal" sizes [28]

Recently, a new model to describe thermotropic copolyesters
in the solid state was advanced by Alan Windle, who suggests that
these materials crystallize by allowing the best-fit sequence matches
between chains to find each other [29]. These three-dimensional
overlaps form the crystalline phase, with "crystals" differing in
chemical composition. This hypothesis is consistent with many of
the observations already described, i.e., "melting," relation of
melting temperature to chemical composition, annealing behavior, etc.

As-spun fibers are generally subjected to a heat treatment
process to improve fiber properties [17,27]. In a typical heat treatment, the fibers are subjected to a high temperature (typically
10-20°C below the melting temperature) and an inert environment
for a prolonged time (10 min to several hours). As a result of heat
treatment, the tenacity shows a dramatic increase from about 10 g/d
for as-spun fibers to values in excess of 20 g/d, occasionally reach-
ing as high as 40 g/d. In addition, the chemical resistance and
the melting point (and therefore the use temperature) also increase.
Most of the thermotropic copolyester fibers do not show a significant
increase in the fiber modulus (typically about 650 g/d) with the
heat treatment, consistent with the absence of orientation changes
noted above. Because of the improvement in the fiber properties,
heat treatment is an integral part of the fiber process. Most of
the discussion of fiber properties describes the behavior of annealed
fibers.

6. PROPERTIES OF THERMOTROPIC COPOLYESTER FIBERS

The tensile moduli of wholly aromatic, thermotropic copolyester
fibers are typically in the range of 400 to 1100 g/d (Table 3.1),
depending on the specific chemical composition. Fibers from polymers
containing m-phenylene moieties show lower modulus values are
observed for fibers from polymers containing the highly linear 4,4'-
biphenylene moiety. The modulus of fibers from 2,6-naphthylene
containing polymers are in the range of 500 to 700 g/d and do not
increase with heat treatment.

Table 3.1   Tensile Modulii of Thermotropic Copolyester Fibers

Composition Mole %	Modulus, g/d
HBA/IA/HQ 35/32.5/32.5	450
PhenylHQ/TA 50/50	490
HBA/HNA 75/25	600
HNA/TA/HQ 60/20/20	680
HBA/DHN/TA 60/20/20	750
HBA/HNA/TA/BP 60/10/15/15	1080

[a]HBA = p-hydroxybenzoic acid; IA = isophthalic acid; HQ = hydroquinone; PhenylHQ = phenyl hydroquinone; TA = terephthalic acid; HNA = 2-hydroxy-6-naphthoic acid; DHN = 2,6-dihydroxynaphthalene; BP = 4,4'-biphenol.

The tenacity of as-prepared fibers are generally in the range of 5-14 g/d and is primarily a function of the molecular weight of the polymer. While systematic variation of the tenacity with the chemical composition of the copolyesters is not observed, the polymers containing m-phenylene moieties generally show lower tenacity levels. The tenacity of heat treated fibers are in the range of 20-40 g/d, and again, no systematic variation of tenacity with the polymer composition is observed.

The tenacity increase with heat treatment is believed to be substantially due to increased molecular weight (solid state polymerization) rather than to the structural perfecting processes occurring. A model of the tenacity of highly oriented, rigid chain fibers has been developed [28]. Figure 18 illustrates the molecular weight dependence of the tenacity of fibers from HBA/HNA copolyesters, with the solid line calculated using the model. Excellent agreement between experiment and theory was obtained.

The stress-strain behavior of thermotropic copolyester fibers as a function of temperature is shown in Fig. 19. Unlike many other

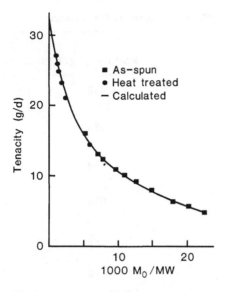

Figure 3.18 Relationship between fiber tenacity and molecular weight of HBA/HNA copolyester fibers.

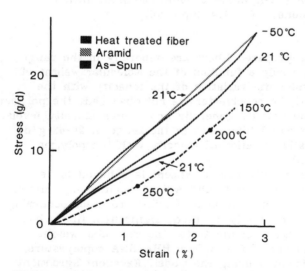

Figure 3.19 Temperature dependence of the stress–strain behavior of a thermotropic copolyester fiber.

high modulus fibers, the fibers do not show linear stress-strain behavior at room temperature. At subambient temperatures the stress-strain behavior is linear, and there is no loss of elongation down to liquid nitrogen temperatures. Above room temperature the fiber exhibits a concave stress-strain curve, presumably due to an increase in chain orientation with deformation. At temperatures above 150°C the breaking strain decreases. This type of stress-strain behavior was observed for nearly all LCP fibers.

Figure 20 details the tenacity-temperature relationship of a thermotropic copolyester fiber in contrast to a conventional fiber such as PET. Note that the LCP fibers retain tensile property up to much higher temperatures, the fiber shown in Fig. 20 retaining about 65% of its ambient temperature tenacity at 150°C. This occurs despite the fact that the fibers are not highly crystalline in the conventional sense and have a glass transition in the range of 100–150°C. The rigid nature, the long persistence length, and the rodlike packing of the aromatic polyester chains prevents them from retracting into less oriented conformations above the glass temperature and

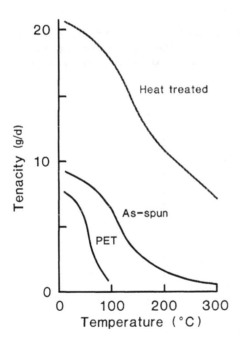

Figure 3.20   Temperature dependence of fiber tenacity of as-spun and heat treated HBA/HNA fibers in comparison to the behavior of PET fiber.

leads to the excellent mechanical property retention noted at high temperatures. The details of the tenacity retention behavior are a strong function of composition and heat treatment conditions.

The temperature dependence of mechanical properties can be clarified by examining the fiber dynamic mechanical behavior. A plot of E' and tan δ as a function of temperature for a HBA/HNA copolymer fiber (11 Hz, Rheovibron) is shown in Fig. 21 [30]. Three loss processes, gamma, beta, and alpha, are observed at -50°C, 41°C, and 110°C, respectively, with changes in E' corresponding to these events. The molecular origins of these processes have been assigned as follows [30]:

> The gamma transition is due to the reorientational motion of p-phenylene groups.
> The beta transition is due to the reorientational motion of 2,6-naphthylene groups. Since the bonds at the 2- and 6-positions are not collinear, this motion requires cooperative motion in neighboring chain units.
> The alpha transition is highly cooperative, similar to a glass transition.

It has been observed that the alpha transition and, to a lesser extent the gamma transition, can be significantly suppressed with annealing. As the beta transition originates from the presence of

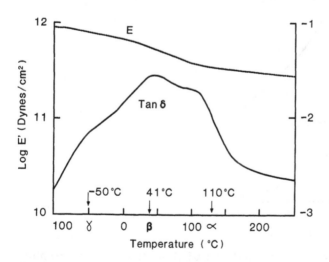

Figure 3.21   Dynamic mechanical behavior of an HBA/HNA copolymer fiber.

Table 3.2 Effects of Solvent Aging on the Mechanical Properties
of a Thermotropic Copolyester and an Aramid Fiber: Percent
Modulus and Strength Retention of Fibers Aged in Solvents for
1 Month at 50°C

Solvent	NTP (5dpf)		Aramid (1.5dpf)	
	% Mod. Ret.	% Str. Ret.	% Mod. Ret.	% Str. Ret.
Original	505	21.4	817	25.6
Water	103	106	95	95
Gasoline	96	95	101	107
Motor oil	96	89	97	92
50% Antifreeze	103	98	99	104
5% Sodium hypochlorite	84	75	Dissolved	
10% Sodium hydroxide	88	40	105	69
20% Sulfuric acid	94	88	101	72
20% Hydrochloric acid	108	117	104	65

2,6-naphthylene units in the polymer chain, it is absent in polymers
that do not contain naphthalenic units.

Fibers of wholly aromatic, thermotropic polymers generally possess
excellent chemical resistance, as shown in Table 2. In organic solvents
and acidic environment, the fiber retain tensile properties for long
periods of exposure. Only under alkaline conditions do the properties
show significant dropoff. The alkaline resistance of these fibers
is a strong function of chemical composition and is significantly
poorer for fibers containing m-phenylene moieties.

It is now becoming evident that the mechanical properties of
thermotropic copolyester fibers are very similar if the measurements
are made at temperatures significantly below the first- and second-
order transition temperatures associated with a given molecular
architecture. Control of transition temperature and the extent of
property loss at temperature is a function of the overall crystallinity
of the system and the degrees of freedom of the molecular chain,
analogous to conventional polymers. Important for these systems
is the necessarily reciprocal relationship between thermal stability
and processing ease. Interestingly, chemical properties such as
hydrolytic stability and solvent resistance also vary with the details
of the molecular backbone. What one may deduce from these observations is that there are likely a number of chemical approaches to
a given property set, and that polymer cost will therefore play an
even more important role than usual in determining product success.

7. CONCLUSIONS

The discoveries and concepts that led to the first demonstrations
of thermotropic copolyester fibers in the laboratory are now about
a decade old, and the products based on this technology are tracking
well on the expected 10- to 15-year, new material commercial reality
timeframe. In many ways, this technology is benefiting from the high
degree of sophisticated science that has been performed in the liquid
crystal polymer area, which has become a strong focus of both academic
and industrial polymer research. The success of fibers based on thermotropic copolyesters will be determined over the next several years and,
in the absence of true tensile and compressive property uniqueness,
cost/performance and ancillary properties such as hydrolytic stability
and cut resistance will play a more important role than usual in determining the commercial acceptance of this approach to high performance fibers.
The competition, in addition to the many options that lead to tractable
thermotropic copolyesters, include the now well-established aramid
fibers and the emerging high performance polythylene fibers.

ACKNOWLEDGMENTS

The authors would like to thank their many colleagues at the Hoechst
Celanese Research Division, the Vectra Business Unit, and Hoechst
Celanese Fibers Group who have contributed their ideas and efforts
toward making this review possible.

REFERENCES

1. R. S. Jones and M. Jaffe, High Performance Aramid Fibers,
   in *High Technology Fibers, Part A* (M. Lewin and J. Preston,
   Eds.), Marcel Dekker, New York, 1985.
2. S. L. Kwolek, U.S. Patent 3,600,350, 1971.
3. See other chapters in this volume.
4. S. G. Cottis, J. Economy, and B. E. Nowak, U.S. Patent
   3,637,595, 1972.
5. S. G. Cottis, J. Economy, and L. C. Wohrer, U.S. Patent
   3,975,487, 1976.
6. H. F. Kuhfuss and W. J. Jackson, Jr., U.S. Patent 3,778,410,
   1973.
7. G. W. Calundann and M. Jaffe, *Proceedings of the Robert A.
   Welch Conferences on Chemical Research*, 26, 247 (1982).
8. G. W. Calundann, *Am. Chem. Soc., Polym. Preprnts.*, 27,
   493 (1986).

9. M. Jaffe, High Modulus Polymers, in *The Encyclopedia of Polymer Science and Engineering*, John Wiley and Sons, New York, to be published.
10. G. Calundann, M. Jaffe, and H. N. Yoon, in *Fibers for Composites* (A. R. Bunsell, Ed.), Elsevier *Science Publishers*, to be published in 1988.
11. A. Cifferri and I. M. Ward., Eds., *Ultra High Modulus Polymers*, Applied Science Publishers, Barking, U.K., 1979.
12. K. F. Wissbrun, *Bri. Polym. J.*, 163 (1980).
13. W. J. Jackson, Jr., and H. F. Kunfuss, *J. Polym. Sci., Polym. Chem. Ed.*, 14, 2043 (1976).
14. *Chem. Corresp.*, October 20, 1984.
15. *Chem. Corresp.*, November 5, 1985.
16. J. J. Kleinschuster, U.S. Patent 3,991,014, 1976; T. C. Pletcher, U.S. Patent 3,991,013, 1976; J. J. Kleinschuster and T. C. Pletcher, U.S. Patent 4,066,620, 1978; J. R. Schaefgen, U.S. Patents 4,075,262 and 4,118,372, 1978; C. R. Payet, U.S. Patent 4,159,365, 1979; R. S. Irwin, U.S. Patents 4,232,143, 4,232,144, 1980 and 4,245,082, 1981.
17. G. W. Calundann, U.S. Patent 4,185,996, 1980.
18. G. W. Calundan, U.S. Patent 4,161,470, 1979.
19. A. J. East, L. F. Charbonneau and G. W. Calundann, U.S. Patent 4,330,457, 1982.
20. I. Hay, private communication.
21. L. C. Sawyer and M. Jaffe, *J. Mater. Sci.*, 1897 (1986).
22. E. Baer, A. Hiltner, T. Weng, L. C. Sawyer, and M. Jaffe, *Polym. Materials Sci. Eng.*, 52, 88 (1985).
23. T. Weng, A. Hiltner, and E. Baer, *J. Mater. Sci.*, 21,744 (1986).
24. G. A. Gutierrez, J. Blackwell, J. B. Stamatoff, and H. N. Yoon, *Polymer*, 24, 937 (1983).
25. J. B. Stamatoff, *Mol. Cryst. Liq. Cryst.*, 110, 75 (1984).
26. J. Blackwell and G. Gutierrez, *Polymer*, 23, 171 (1982).
27. H. N. Yoon, private communication.
28. H. N. Yoon, to be published.
29. A. Windle, to be published.
30. H. N. Yoon and M. Jaffe, to be published.
31. R. A. Chivers, J. Blackwell, G. A. Gutierrez, J. B. Stamatoff, and H. N. Yoon, in *Polymer Liquid Crystals* (A. Blumstein, Ed.), Plenum Press, New York, 1985, pp. 153-166.

4

POLYMER SINGLE CRYSTAL FIBERS

ROBERT J. YOUNG / Manchester Materials Science Centre,
Manchester, England

1. INTRODUCTION

In the past, most of the effort in obtaining highly oriented polymer samples has been concentrated on trying to uncoil and align long polymer molecules in the solid state, melt, or solution. A wide variety of techniques have been employed with conventional polymers such as polyethylene. These have included ultra-drawing [1-3], high degrees of extrusion [3-5], and gel or solution spinning [6,7]. Also, new, rigid rod polymers have been developed from which highly oriented fibers can be prepared by the spinning of liquid crystalline solutions [8-10]. Significant improvements and developments have taken place over recent years, and samples with high values of modulus and strength have been prepared. However, perfect molecular alignment as in single crystals is never achieved using such approaches, since defects such as chain ends, chain folds, loops, and entanglements are invariably trapped in the structures.

The technique of solid state polymerization has the ability to produce polymer single crystal fibers in which the molecules are perfectly aligned and the fibers have their theoretical values of stiffness and strength. Solid state polymerization is essentially very simple. Single crystals of monomer molecules are prepared using conventional methods such as vapor phase deposition or precipitation from dilute solution. Alignment of the monomer molecules is present in the monomer crystals, and the transformation from monomer to polymer takes place through a solid state polymerization reaction by a rearrangement of bonding within crystals but without any appreciable molecular movement. Because of this, the alignment of the monomer molecules is transferred to a uniaxial orientation of the polymer molecules. The normal problems encountered with aligning polymer molecules in the melt or solution, such as entanglements, are therefore avoided. In this way polymer molecules can be incorporated into fibers without passing through the melt or solution stage. This invariably produces polycrystalline samples and imperfections in the fibers.

Even though the solid state polymerization technique is simple and attractive, it suffers from several drawbacks. The fibers, which are produced relatively slowly, tend to be short strands with maximum lengths of no more than 5-10 cm. In addition, the only successful single crystal fiber system is based on the solid state polymerization of certain substituted diacetylenes [11-13], and it cannot be applied to all polymer systems. In spite of this, these substituted diacetylene polymer fibers are found to have promising and useful physical properties characteristic of one-dimensional solid. Even though this chapter is concerned principally with the mechanical properties of polydiacetylene single crystals, it should be remembered that they also have unusual optical [14,15] and electronic properties [16,17] and have recently been termed "prototype one-dimensional semiconductors" [18]. The production and exploitation of polydiacetylene single crystals has led to significant improvements in our understanding of many structure/property relationships in these one-dimensional solids.

2. SOLID-STATE POLYMERIZATION REACTIONS

2.1 Types of Reactions

The different types of solid state polymerization reactions have been discussed in detail by Wegner [11], and his terminology has been adopted in this chapter. The solid state reactions that are relevant to the preparation of polymer single crystal fibers are termed either "topochemical" or "topotactic," depending on the detailed nature of the reaction. These terms are used to describe reactions that can take place in organic crystals such as single crystals of fiber-forming monomers. In such monomer crystals the molecules are generally separated sufficiently far that they are not able to react. However, if there is sufficient mobility that the molecules are able to move to within about 0.3 nm of each other by diffusion or rotation, then a solid state reaction may occur. Examples [19-21] of such systems in which monomer single crystals can undergo solid state polymerization have been given by Wegner [11]. The most important reactions for the formation of good polymer single crystals are topochemical reactions whereby, as Wegner [11] pointed out, "there is a direct transition from the monomer molecules to polymer chains without destruction of the crystal lattice and without the formation of non-crystalline intermediates."

The behavior is therefore rather like a martensitic transformation [22]. It is illustrated in Fig. 1 for the solid state polymerization of a substituted diacetylene. It can be seen that the centers of gravity of the molecules do not move significantly from their lattice site, and the reaction takes place within the parent crystal. The three-dimensional order of the monomer lattice is therefore transferred directly to the polymer lattice, with the result that the polymer formed must have a well-ordered extended chain morphology. This type of reaction was first reported in 1969 by Wegner [12], who produced virtually defect free, extended chain, single crystals of substituted polydiacetylenes from single crystal monomers. Since 1969 a large number of different substituted diacetylenes have been produced, and a number of examples are shown in Table 1 along with the abbreviations that are used to describe them in this chapter.

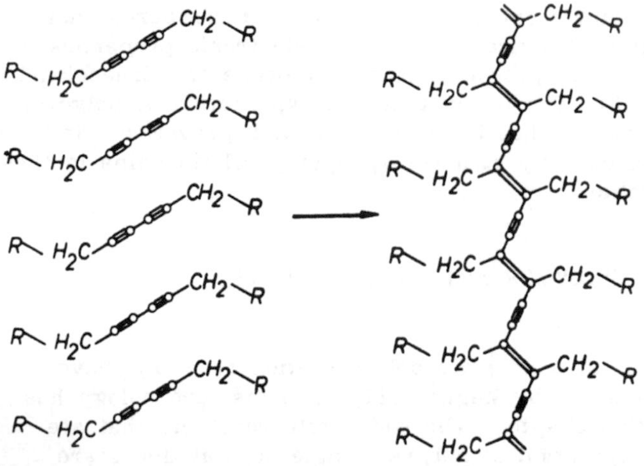

Figure 4.1 Topochemical solid state polymerization reaction of a disubstituted diacetylene. Reproduced from Ref. 11 with permission.

Table 1 Formulas and Abbreviations for the Diacetylene Derivatives Described in the Text with General Formula $R-CH_2-C\equiv C-C\equiv C-CH_2-R$

R	Chemical name	Abbreviation
$-OCONHC_2H_5$	2,4-Hexadiyne-1,6-diol-bis(ethyl urethane)	EUHD
$-OCONHC_6H_5$	2,4-Hexadiyne-1,6-diol-bis(phenyl urethane)	PUHD
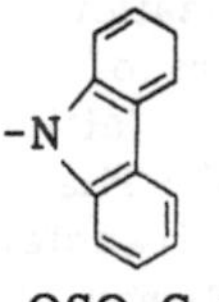 $-N$ (carbazole ring)	2,4-Hexadiyne-1,6-di(N-carbazol)	DCHD
$-OSO_2C_6H_4CH_3$	2,4-Hexadiyne-1,6-diol-bis(p-toluene sulfonate)	TSHD

2.2   Mechanisms and Kinetics of the Reaction

Topochemical polymerization reactions can proceed in either a homogeneous or heterogeneous manner as shown in Fig. 2. In the homogeneous case (Fig. 2a), the polymer molecules grow randomly in the monomer matrix. The other possibility is that the polymerization is heterogeneous (Fig. 2b), starting at specific sites in the monomer crystal such as at defects. When further chain growth takes place at the preexisting polymer nuclei, there can be a tendency for the polymers formed to be polycrystalline. This is found to occur in four center-type polymerization [21]. Both homogeneous and heterogeneous polymerizations have been reported in the case of polydi-

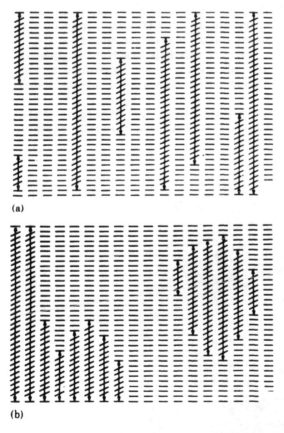

(a)

(b)

Figure 4.2   Two possible methods of forming polymer molecules during solid state polymerization: (a) homogeneous reaction, (b) heterogeneous reaction. (Reproduced from Ref. 11 with permission.)

acetylenes. When the toluene sulfonate derivative (TSHD) is polymerized using synchrotron radiation, the reaction has been observed
to take place homogeneously throughout a crystal [23]; whereas
when TSHD monomer is polymerized thermally, the polymerization
reaction is found to start preferentially at defects [24]. In both
cases the polymer forms as a solid solution with monomer, and the
single crystal morphology is directly transferred from the monomer
to the polymer, leading to poly-TSHD crystals having a high degree
of perfection. This perfection can be clearly seen in the rotation
photograph in Fig. 3.

The polymerization behavior of other diacetylene derivatives
can be more complex, such as in the case of fiber-forming diacetylenes
such as the carbazolyl (DCHD) [25,26] and ethyl urethane (EUHD)
[27-29] derivatives. In both cases thermal polymerization produces
polycrystalline samples that are not as crystallographically perfect
as the single crystals produced when polymerized using γ-rays
at room temperature. In the case of DHCD it has been suggested
that this may be due to the tendency for phase separation to be
favored kinetically at the higher temperatures used for the thermal
polymerization.

Several different mechanisms have been put forward to explain
the formation of polymer chains [11]. The most generally accepted

Figure 4.3 An X-ray rotation photograph of a poly-TSHD single
crystal. The chain direction was the rotation axis.

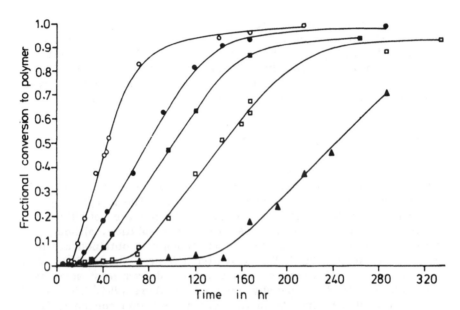

Figure 4.4   Fractional conversion as a function of time for the thermal polymerization of EUHD at different temperatures: ▲ -70°C, □ -75°C, ■ -78°C, ● -80°C, ○ -86°C. (Reproduced from Ref. 27 with permission.)

one uses evidence from electron spin resonance spectroscopy [30]. Using this technique it is found that the reactive species is a carbene and not a free radical as was first expected. There have also been several attempts to model the reaction kinetics, and the most successful attempt has been made by Baughman [31]. Conversion-time curves for the polymerization of many diacetylenes have characteristic sigmoidal shapes, as can be seen in Fig. 4 for the thermal polymerization of EUHD. The polymerization of most diacetylenes involves a small change in crystal lattice parameters (~5%), and this causes internal strains as the polymer chains grow inside the monomer lattice. This behavior and the shape of the curves has been explained successfully in terms of crystal strain models [31]. However, the fact that it also works for polymerizing polydiacetylene systems that are not single phase raises some doubt about the generality of the model and also suggests that some of the fits that have been reported in the literature [27] may be fortuitous.

As the high modulus, fiber forming diacetylenes are insoluble once they have been polymerized, it is impossible to determine the molar mass or degree of polymerization of the polymers. The crystal fibers behave as if the polymer molar mass is effectively infinite, although recent work on soluble diacetylene derivatives with long aliphatic side groups [32] has suggested that solid state polymerization leads to degrees of polymerization of the order of 1000-1500. However, it is a matter of debate as to how relevant this observation is to the high modulus derivatives with short side groups.

3. FIBER MORPHOLOGY AND STRUCTURE

3.1 Morphology

Polydiacetylene single crystals can be obtained essentially in two crystal forms, either as lozenges or as fibers. The morphology is controlled by both the R group (Table 1) and the conditions under which the monomer is crystallized from solution, although the exact reasons why a particular morphology is obtained are not really understood. The TSHD derivative (Table 1) is normally found only in the form of lozenges when crystallized from most solvents [13], whereas DCHD is usually obtained as fibers [25]. The aspect ratio of the fibers depends on the solvent, solution concentration, and crystallization temperature. EUHD, in contrast, can be obtained in three crystal forms, one of which can undergo solid state polymerization to give fully polymerized single crystal fibers [27], an example of which is shown in Fig. 5. It is found that good polymerizable fibers of PUHD (Table 1) are obtained when it is crystallized from dioxane/water mixtures and dioxane molecules are trapped interstitially in the structure [33]. The crystal morphology in each of these cases is retained when the monomer crystals are transformed into polymer by the solid state polymerization reaction.

A very convenient way of investigating the structure of polydiacetylenes is by transmission electron microscopy. Monomer crystals that are sufficiently thin (typically 100 nm) to allow penetration of the electron beam can be readily produced by allowing a droplet of dilute monomer solution to evaporate on a carbon support film on an electron microscope grid [34]. Polymerization of the monomer can then be achieved by heating or, most conveniently, by exposure to the electron beam in the microscope.

3.2 Structure

The crystal structures of several polydiacetylenes have been determined to a high degree of accuracy using X-ray diffraction methods [35-37]. This should be contrasted with conventional polymers, for which the crystal structures [38] determined using oriented polycrystalline samples are much less accurately known. The perfect

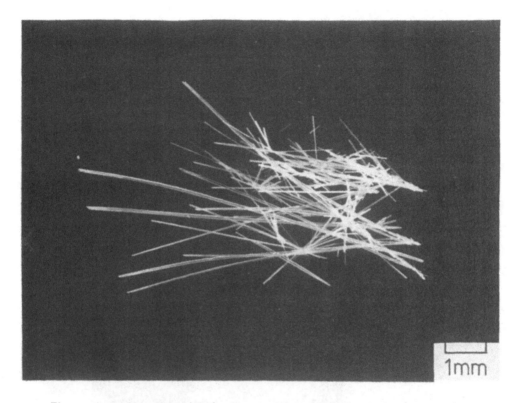

Figure 4.5 Macroscopic single crystals of EUHD. (Reproduced from Ref. 29 with permission.)

single crystal nature of polydiacetylenes has allowed the positions of the atoms in the unit cells and bond angles to be measured to a high degree of accuracy and precision. The monomers and polymers of diacetylenes generally have monoclinic P2$_1$/c crystal structures (where the chain direction is indexed unconventionally as b). The unit cell parameters of several common monomers and polymers are given in Table 2, and this detailed knowledge of the crystal structures is invaluable for the correlation of physical properties with structure in these polymers.

Even though X-ray diffraction is by far the most accurate method of determining crystal structures, extra information can be obtained using electron microscopy. This also has the advantage of producing high magnification images of the structure and performing simultaneous electron diffraction. Single crystal fibers of poly-DCHD are shown in Fig. 6 with a diffraction pattern obtained from one of the

Table 4.2 Details of the Unit Cells of the Polymers of the Four
Common Diacetylene Derivatives from Table 1; Chain Direction Is
the b axis

Derivative	a/nm	b/nm	c/nm	β(deg)	Ref.
EUHD	1.797	0.488	1.514	98.4	27
PUHD[a]	0.489	1.253	1.677	96.8	36
DCHD	1.287	0.491	1.740	108.3	37
TSHD	1.449	0.491	1.494	118.0	35

[a]Triclinic (P$\bar{1}$) with a as the chain direction and α = 69.3° and
γ = 96.2°.

(a)

Figure 4.6 Single crystal fibers of poly-DCHD viewed in a transmission electron microscope: (a) various crystals and typical diffraction pattern (inset), (b) high resolution lattice image shown a spacing
of 1.2 nm of planes parallel to the chain direction. (Reproduced
from Ref. 26 with permission of the publishers, John Wiley & Sons.)

crystals. The diffraction pattern confirms that the chain direction is parallel to the fiber axis. Also, the pattern shows that the polymer fibers are true single crystals with no amorphous scattering.

A problem normally encountered with the study of organic materials in the electron microscope is radiation damage, whereby the crystal structure is destroyed by exposure to the electron beam [39]. It is found that most polydiacetylenes are relatively stable, but that poly-DCHD is particularly good in its ability to resist damage in the electron beam, being at least 20 times more stable than polyethylene [40], the most widely studied crystalline polymer. The high radiation stability has allowed detailed investigations to be made of the structure of poly-DCHD at high magnification. It has been possible to image the crystal lattice directly in the microscope [40,41], as is shown in Fig. 6b, where lattice planes parallel to the chain direction are imaged. These observations indicate that there is perfect molecular alignment and a high degree of crystal perfection in the fibers. In contrast, highly defective polycrystalline structures are found during similar electron microscope observations of polymer fibers produced using techniques such as mechanical orientation [42] and solvent spinning of liquid crystalline solutions of rigid rod polymers [43]. The difference in structure is reflected

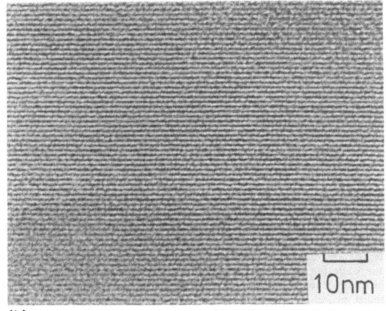

10nm

(b)

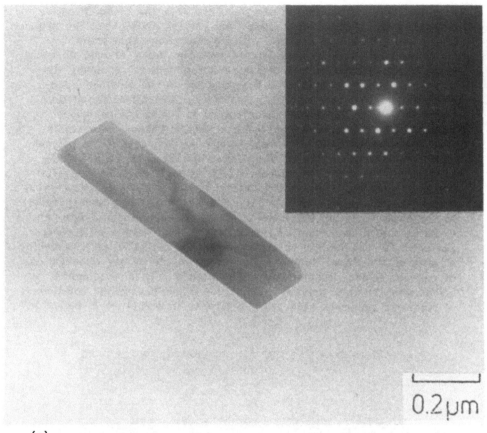

(a)

Table 4.7 Lamellar single crystals of poly-DCHD with molecules perpendicular to the surface: (a) electron micrograph and electron diffraction pattern (inset), (b) filtered high-resolution micrograph showing individual molecules and simulated image (inset). (Reproduced from Ref. 44 with permission of the publishers.)

directly in the differences in the mechanical properties between the single crystal fibers and fibers produced using other techniques, as will be discussed later.

An important factor that exerts control on the Young's modulus of polymer crystals is the sideways packing of the molecules. This can be elucidated by viewing the crystals parallel to the chain direction. Recent studies [44,45] have shown that it is possible to prepare lamellar crystals of poly-DCHD (Fig. 7a) in which the molecules are

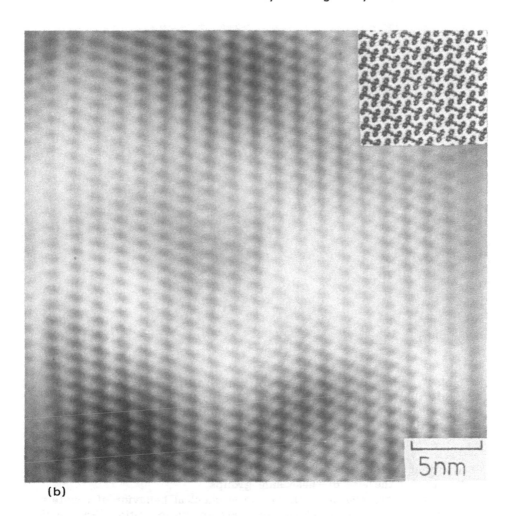

(b)

perpendicular to the crystal surface and so parallel to the electron
beam. Electron micrographs obtained at high magnification show
a full structural image, which can be enhanced by removing any
noise using conventional techniques [46]. Figure 7b shows an enhanced molecular image of poly-DCHD crystal viewed parallel to
the molecules. The projections of the side groups overlap so that
the profiles of the molecules can be seen. Also in the inset is a
computer-simulated image of the structure, which has been obtained
from knowledge of the crystal structure [37] of poly-DCHD. It
can be seen that the molecules consist of a small backbone with
two relatively large, dumbbell-shaped side groups. It is shown

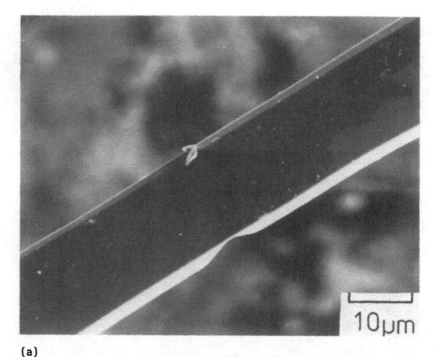

(a)

Table 4.8 Steps on the surface of poly-DCHD crystals: (a) macroscopic step seen in SEM (reproduced from Ref. 26 with permission), (b) molecular steps seen in a high resolution lattice image (micrograph supplied by P. H. J. Yeung).

later that Young's moduli of polydiacetylene fibers are controlled directly by the size of these side groups.

Defects are known to affect the mechanical behavior of crystals of most materials, and polymer crystals are no exception. Preparation of the chain folded, solution grown, lamellar single crystals of conventional polymers prepared from dilute solution have allowed limited investigations to be carried out on certain types of dislocations [47-49]. Polydiacetylene single crystals have now allowed detailed studies of several types of defects found in polymer single crystals to be carried out because of their good stability in the microscope, their well-defined structure, and the ability to view crystals at different angles to the chain direction.

Detailed studies on defects in polydiacetylene single crystals have been carried out on poly-TSHD, and edge dislocations with Burgers vectors parallel to the chain direction have been reported in crystals of this material [50,51]. The density of these dislocations is found to be of the order of 10^{13} m^{-2}. In some cases they can

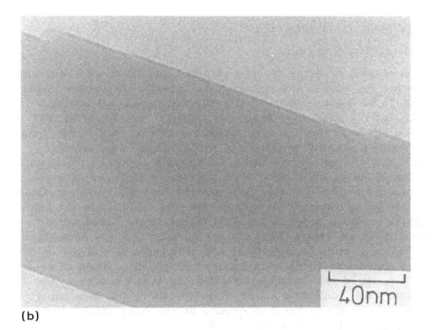

(b)

form a small-angle grain boundary [50]. Edge dislocations with
Burgers vectors perpendicular to the chain direction consisting
of an extra half plane of chain ends have been imaged in poly-DCHD
fibers using high resolution techniques [40]. It is most likely that
both of these types of dislocations are present initially in the monomer crystals and become frozen into the polymer during polymerization. Although both types of dislocations are probably capable of
moving in the monomer, they cannot move in the polymer without
breaking covalent bonds and so are unlikely to be involved in deformation. In addition, stacking faults have also been seen in crystals
of poly-TSHD [52].

The presence of surface defects such as steps is known to
have an important effect on the fracture behavior of polydiacetylene
single crystal fibers. The geometry of such defects has been examined
in detail [26], and examples of steps on the surfaces of poly-DCHD
fibers are given in Fig. 8. It is likely that the steps form during
growth of the monomer crystals from solution as with dislocations.
In microscopic fibers such as the one shown in Fig. 8b, it can
be seen that crystal growth has taken place by the deposition of
individual molecular layers. Even though the two fibers have considerably different diameters, the size of the steps in both cases
is of the order of 1/10th to 1/20th of the fiber diameter. It will
be shown later that this has an important effect on the size dependence of the strengths of the crystals.

4. MECHANICAL PROPERTIES

4.1 Stress-Strain Curves

The preparation of single crystals of polydiacetylenes with centimeter dimensions has allowed the stress-strain behavior of polymer single crystals to be determined using conventional mechanical testing methods for the first time. Baughman, Gleiter, and Sendfeld [33] first demonstrated that fiberlike single crystals of poly-PUHD could be deformed elastically to strains in excess of 3%. The polymer crystals were found to have values of Young's moduli in the chain direction of the order 45 GPa. High degrees of stiffness have also been found for other polydiacetylenes, with Young's modulus values of 45 GPa being reported for poly-DCHD [26] and 62 GPa for poly-EUHD [29]. Figure 9a gives a typical stress-strain curve for a poly-DCHD single crystal fiber. It can be seen that the curve is linear up to a strain of about 1.8% and that there is a slight decrease in slope above this strain until fracture occurs at a strain of about 2.8%. It is found that loading and unloading take place along the same path [26], indicating a lack of hysteresis in the deformation. The fracture strain is found to depend on the fiber diameter, decreasing as the diameter increases [26].

The stress-strain curve in Fig. 9a gives a clear indication of how deformation takes place on the molecular level, since it corresponds essentially to a molecular stress-strain curve. It has been suggested [33] that the deviation from Hooke's law above about 2% strain, which is not a yield process, might be due to the anharmonic part of the interaction potential between neighboring atoms on the polymer chain. As the polydiacetylene fibers are highly perfect polymer single crystals, the deformation directly involves the stretching and bending of bonds along the polymer backbone. This has been demonstrated using resonance Raman spectroscopy, where it has been shown that the frequencies of the C-C, C=C, and C≡C stretching modes in polydiacetylenes depend on the deformation of the crystals and decrease with applied strain. Figure 9b shows an example of the variation of the frequency of the C≡C stretching mode for poly-DCHD as a function of applied strain. The consequent reduction in force constants is one of the main factors leading to the reduction in the slope of the stress-strain curves that is found at high strains.

4.2 Young's Modulus

The levels of chain direction modulus displayed by polydiacetylene single crystals are very high when account was taken of the high cross-sectional area of the molecular chains (A) caused by the

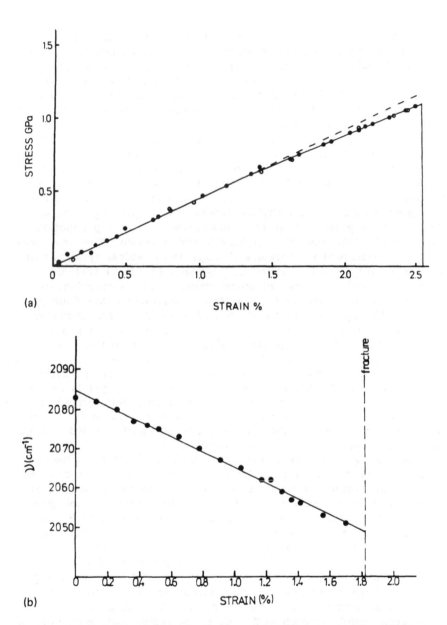

Figure 4.9 Deformation of single crystal fibers of poly-DCHD:
(a) stress-strain curve for both loading (●) and unloading (○)
(reproduced from Ref. 28 with permission), (b) dependence of
C≡C Raman stretching frequency upon strain.

relatively large side groups on the polydiacetylene molecules. In
Fig. 10a the chain direction modulus (E) of various polydiacetylenes
is plotted against the reciprocal of the area supported by each
chain in the crystal, 1/A, which was determined from the knowledge
of the crystal structures (Table 2). The data fall close to a straight
line, demonstrating clearly the direct relationship between the modulus
and crystal structure. The value of 1/A for polyethylene is also
indicated in Fig. 10a, and assuming that the two backbones have
about the same stiffness, then a modulus of the order of 250 GPa
is predicted for polyethylene. This value is in line with current
theoretical predictions [55]. The fact that this per-chain modulus
is close to that of diamond [56] has led to many efforts being made
to align molecules in polyethylene fibers [1-7]. The alignment in
the case of polydiacetylenes is a consequence of their production
by a topochemical solid state polymerization reaction. The way forward
with these materials is therefore through the chemical synthesis of
monomers with smaller substituent R groups (Table 1) that are
capable of forming fibers and polymerizing to full conversion. It
is found that for a particular diacetylene derivatives the Young's
modulus (E) depends on both degree of conversion from monomer
to polymer and the method of polymerization. Figure 10b shows
the dependence of E on the conversion of EUHD to polymer using
both heat and γ-rays [29]. There is a linear increase in the modulus
with the volume fraction of polymer in the crystals for the thermal
polymerization of EUHD, whereas the moduli of the crystals produced
by γ-ray polymerization are lower. The thermally polymerized crystals
are known to be crystallographically less perfect, but it is thought
that the high doses of radiation needed to polymerize EUHD damage
the crystals, giving rise to lower values of E.

The Young's modulus in the chain direction is of the greatest
importance for engineering applications, but it must be remembered
that a large number of elastic constants are needed to fully describe
the elastic behavior of a crystal [57]. Single crystals of polydiacetylenes usually possess monoclinic symmetry and so have 13 elastic
constants. Several attempts have been made to calculate the nine
elastic constants for orthorhombic polyethylene [58], but there
appear to have been no similar calculations for a polydiacetylene.
This calculation should be feasible, as the crystal structures and
atomic positions of several polydiacetylenes are known to a high
degree of accuracy [35-37]; but there are also a large number
of different bonds in each unit cell, which makes such a calculation
extremely complex.

Batchelder and Bloor [53] used the method of Treloar [59]
to calculate the modulus of poly-TSHD crystals deformed in the
chain direction. They employed force constants measured from

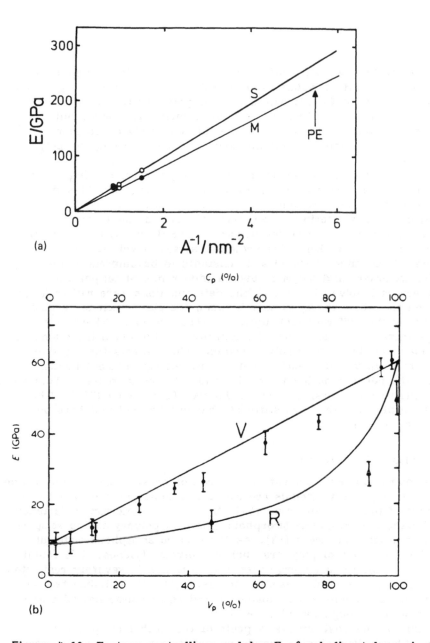

Figure 4.10 Factors controlling modulus E of polydiacetylene single crystal fibers: (a) Variation of E determined both spectroscopically (S) and mechanically (M) with reciprocal of area supported by each chain of the crystals. The arrow indicates the value of 1/A for polyethylene. (Reproduced from Ref. 26 with permission.) (b) Dependence of E on polymer volume fraction V_p and conversion to polymer C_p for thermally polymerized (●) and γ-ray polymerized (△) EUHD fibers. (Reproduced from Ref. 29 with permission.)

Raman spectroscopy and estimated the chain direction Young's modulus for poly-TSHD crystals to be 50 GPa, which is in good agreement
with the value of 45 GPa determined for poly-DCHD [26] with similar
unit cell dimensions. Treloar's method is relatively simple, but it
cannot be extended to determine the full set of elastic constants
as it only takes into account the covalent bonding along the polymer
backbone [59].

Some of the 13 elastic constants for polydiacetylene single crystals have been measured using Brillouin scattering [60,61] and
sound velocity measurements [62]. Leyrer and co-workers [60]
used Brillouin scattering to determine six of the elastic constants
of the monomer and three for polymer of TSHD at room temperature.
Rehwald and co-workers [62] showed that sound velocity measurements allowed nine of the elastic constants to be determined for
TSHD monomers and polymers over a wide range of temperature.
The errors involved in some of the determinations were rather large,
but on the whole the agreement between the elastic constants determined by the different techniques is relatively good. There is a
large increase, by an order of magnitude, in the chain direction
Young's modulus during polymerization. This reflects the replacement of weak van der Waals bonding with strong covalent bonds
along the polymer backbone and is similar to the increase in Young's
modulus with conversion shown in Figure 10b for poly-EUHD. This
again emphasizes the high stiffness that can be achieved from polymer molecules [56].

4.3 Plastic Deformation

The restriction that covalent bonds are not broken during deformation
means that polymer crystals are capable of undergoing only a limited
amount of plastic deformation [63]. Much of the deformation is taken
up by the noncrystalline amorphous phase in polycrystalline samples
of conventional polymers [63]. As there is no amorphous material
in single crystals of polymers such as polydiacetylenes, it is found
that their ability to undergo plastic deformation is severely restricted
and so they are highly resistant to creep [26,29]. Polydiacetylenes,
however, are found to be capable of undergoing some limited plastic
deformation through twinning [64,65].

One of the most unusual aspects of the mechanical properties
of polydiacetylene single crystals is that it is not possible to measure
any time-dependent deformation such as creep when crystals are
deformed in tension parallel to the chain direction [26,29]. The
lack of creep is shown in Fig. 11 for a poly-DCHD single crystal
held at constant stress at room temperature, and further measurements have indicated that creep could not be detected during

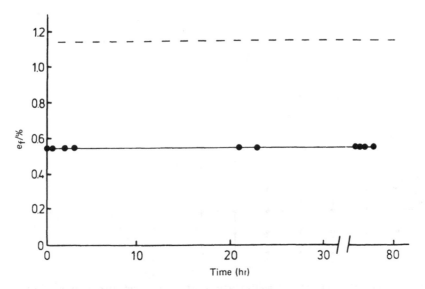

Figure 4.11 Variation of fiber strain with time for a single crystal fiber of poly-DCHD held at a constant stress that corresponded to about half of its fracture strain. (Reproduced from Ref. 26 with permission.)

deformation at temperatures of up to at least 150°C [26]. Creep and time-dependent deformation are normally a serious problem in the use of high-modulus polymer fibers made from flexible molecules such as polyethylene [66]. Oriented polyethylene fibers produced by drawing or spinning contain a high density of defects such as chain ends, loops, and entanglements, and these allow the translation of molecules parallel to the chain direction during deformation, which leads to creep. However, recent work by Ward and co-workers [67] has shown that the creep in such fibers can be dramatically reduced by radiation cross-linking. Polydiacetylene single crystal fibers contain perfectly aligned, long polymer molecules, and so there is no mechanism whereby creep can take place even at high temperatures.

Recent studies on the deformation of polydiacetylene single crystals [64,65] have demonstrated clezrly that polymer single crystals twin when deformed in compression parallel to the chain direction. This type of deformation leads to the formation of twins, which involve the molecules kinking over at a well-defined angle [64,65]. Figure 12 shows a schematic diagram of a twin in a poly-DCHD [68] fiber and the corresponding molecular displacements involved. Such a process can be differentiated from the formation of kink

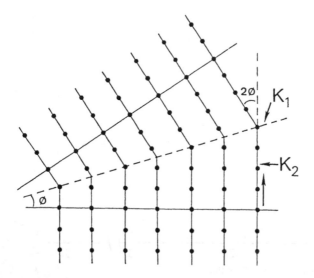

Figure 4.12   Schematic representation of chain twinning in a polydiacetylene single crystal fiber.

bands, which are found in other high modulus fibers [69], since
there is a mirror image orientation relationship between the deformed
and undeformed regions [64], which is not the case with kink bands
[69]. It was previously thought that twinning in polymer crystals
would not be able to take place by the bending of the molecular
chains [70]. However, Pietralla [71] postulated that this type of
twinning might occur in semicrystalline polymers, and Bevis [72]
subsequently performed a detailed analysis of this "chain twinning"
for polyethylene crystals. These present investigations into deformed
crystals of polydiacetylenes have provided the first direct evidence
for chain twinning taking place, but inspection of some of the micrographs of chain-extended polyethylene lamellae [73] has demonstrated
that the twins were probably also present in such materials.
   Chain twinning in polymer single crystal fibers has important
consequences for their use in composites. For example, it is possible
to tie knots in poly-DCHD fibers [26], as can be seen in Fig. 13.
This involves a high degree of deformation, but the crystals cope
with the high strains by undergoing chain twinning on their inside
surfaces, which are subject to compression in the chain direction.
It can be seen in Fig. 13 that there is some cracking parallel to
the chain direction, but the crystals remain relatively intact even
when pulled into a tight knot [26]. This has implications for the
use of the fibers in applications such as composites. Their ability

to absorb strain by twinning means that the fibers should not break up and undergo "fiber attrition" as easily as other fibers such as those of glass and carbon. Recent investigations have shown that polydiacetylene single crystal fibers tend to twin during fabrication in composites when thermosetting matrices such as epoxy resins are employed [68]. The occurrence of resin shrinkage during both curing and cooling from the cure temperature imposes a compressive stress parallel to the fiber axis that can twin the crystals. This is discussed further in Sec. 5.

4.4 Fracture

Polydiacetylene single crystals readily undergo cleavage parallel to the chain direction. This reflects the relative strength of covalent bonding compared with the van der Waals bonding between the polymer molecules. The cleavage takes place preferentially in poly-diacetylenes on certain crystallographic planes [74]. However, the most interesting aspect of the fracture behavior of polymer single

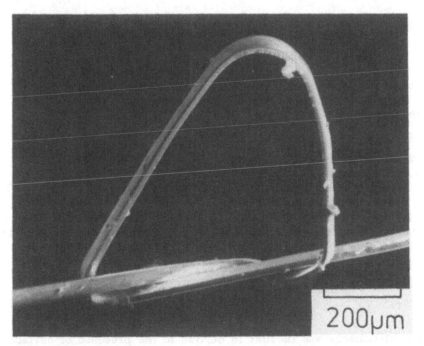

Figure 4.13 Scanning electron micrograph of a knot in a poly-DCHD single crystal fiber.

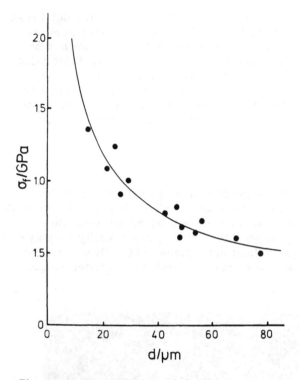

Figure 4.14 Variation of fracture strength with fiber diameter for a poly-DCHD single crystal fiber. (Reproduced from Ref. 26 with permission.)

crystal fibers is the high strength they can exhibit when they are deformed parallel to the chain direction.

Research into the fracture of polydiacetylene single crystal fibers revealed a strong dependence of the fracture stress, σ_f, on the fiber diameter, d [26,29,33]. This is demonstrated for poly-DCHD fibers in Fig. 14. It is similar to the size dependence reported for inorganic high strength fibers [75,76], where the dependence of σ_f on d was found to follow a relation of the form

$$\sigma_f \propto \frac{1}{d} \qquad (1)$$

This relationship was thought to be due to the presence of surface defects, which give rise to a stress concentration when the fibers are deformed. The size of the defects in the inorganic fibers was found to scale with the fiber diameter and so the size dependence of the strength followed [77]. A more detailed examination of the data in Fig. 14 reveals that Eq. (1) is not obeyed accurately for polydiacetylene single crystal fibers. Also, theoretical calculations have shown that a different relationship is expected [26,29]. When the data in Fig. 14 were replotted in the form of a log/log plot, it was found that the dependence of σ_f on d was given more accurately by

$$\sigma_f \propto \frac{1}{d^{0.55}} \tag{2}$$

which is close to the dependence predicted from the theoretical consideration. Detailed analysis of defects (Fig. 8) has allowed the theoretical strength of poly-DCHD crystals to be calculated as 3 ± 1 GPa [26].

It was pointed out by Frank [56] several years ago that polymer molecules can have very high values of strength when deformed parallel to their length. This determination of the theoretical strength of polydiacetylene single crystal fibers has allowed the strength of individual molecules to be estimated. It can be shown from the knowledge of the crystal structure of poly-DCHD [37] that each molecule supports a cross-sectional area A of the order of 1 nm^2. A theoretical strength of 3 GPa therefore corresponds to a force to break molecules of about 3 nN and a fracture strain of about 6-8%. It is of interest to compare this estimate with the theoretically calculated values of strengths of covalently bonded polymer molecules [76,78,79]. Kelly [76] estimated that the strength of a polyethylene molecule is about 6 nN, but this is now thought to be on the high side. Kausch [79] showed that a covalently bonded polymer molecule should be broken by a force of about 3 nN, a value that is identical to that determined for poly-DCHD. This magnitude of molecular strength corresponds to a fracture stress of the order of 20 GPa for a polyethylene single crystal. However, in this case the area supported by each molecule is considerably smaller than for poly-DCHD. Unfortunately, it has not yet been possible to make single crystals of polyethylene with macroscopic dimensions, and so even highly oriented polyethylene fibers have strength values only of the order of 4 GPa [3,5], well below the theoretical strength of polyethylene.

5. COMPOSITES

Composites produced by incorporating high modulus fibers in a brittle matrix such as an epoxy resin are known to have outstanding mechanical properties [53]. Polymeric examples include composites produced with high modulus polyethylene fibers [81] and aromatic polyamide fibers such as Kevlar [82]. Polydiacetylene single crystal fibers are attractive as reinforcing fibers in polymer matrices because of their good mechanical properties, excellent thermal stability, and low density. Investigations into the behavior of polydiacetylene single crystal fibers in epoxy resin matrices [68,83,84] have shown that not only do such composites have promising mechanical properties, but they behave as excellent model systems, allowing the fundamental details of the mechanisms of fiber reinforcement to be revealed.

5.1 Composite Micromechanics

It was shown in Fig. 7b that the variation frequencies of certain main chain, Raman active modes were found to be dependent on the level of applied strain [28,53,54]. In particular, it was found that the C$\equiv$C triple bond stretching frequency changes by the order of 20 cm^{-1} for 1% of strain [53]. This phenomenon can be used to determine the strain in a polydiacetylene fiber subjected to any general state of stress. The strain can be measured to a high degree of spatial resolution and accuracy, as beam diameters of as small as 1 μm can be employed using a Raman microprobe, and changes in frequency can be determined to better than ±1 cm^{-1}. This means that the fiber behaves as though it possesses an internal molecular strain gauge. The Raman strain measurement technique has been recently employed to monitor the point-to-point variation in strain in polydiacetylene fibers in epoxy composites [68,83,84].

An interesting model system is that of a single short fiber in an infinite matrix subjected to an overall strain. This is a classical problem in the theory of fiber reinforcement [80,85]. Model specimens were prepared consisting of a single polydiacetylene fiber in an epoxy bar as shown in Fig. 15a. The matrix strain can be measured using a strain gauge on the surface of the specimen and the point-to-point variation in strain monitored using Raman spectroscopy. Typical results [83] are in Fig. 15b, where the fiber strain is plotted as a function of position along the fiber for different levels of applied matrix strain. It was found that the results agreed qualitatively with theoretical predictions of Cox [85]. For example, at higher levels of matrix strain the fiber strain rises from the end to a constant value along the length of the fiber and then falls off at the other end as predicted by the theory. These regions of rise and fall of strain are known as the "critical length" [80]. It was also demonstrated [83] that the critical length was proportional to the fiber diameter. This result was expected from the theoretical analyses [83,85], but this was the first time it has been demonstrated by direct experiment for a composite system.

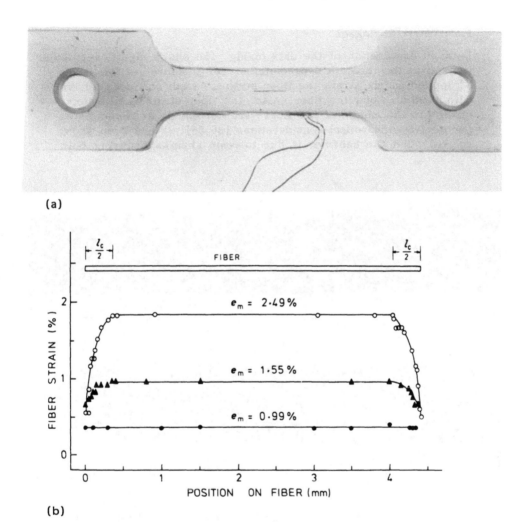

(a)

(b)

Figure 4.15 Polydiacetylene/epoxy model composites: (a) specimen containing a single poly-DCHD fiber, (b) variation of fiber strain with position for different stated levels of matrix strain e_m. (Reproduced from Ref. 83 with permission.)

5.2 Matrix Shrinkage

Detailed examination of the data in Fig. 15b shows significant devia-
tions from the theoretical predictions. For instance, at low levels
of applied matrix strain the fiber strain is small and does not vary
with position along the fiber. Also, the fiber strains and matrix
strains in the middle of the fibers are not equal, as would be expected from theoretical considerations [80,85]. It has been shown
recently that the behavior is due to resin shrinkage during both

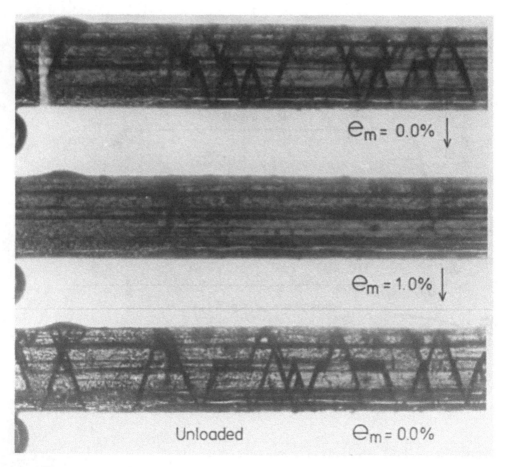

Figure 4.16   Micrographs of a fiber inside a model composite with
an epoxy resin matrix cured at 100°C, taken at different levels
of matrix strain. (Supplied by I. M. Robinson.)

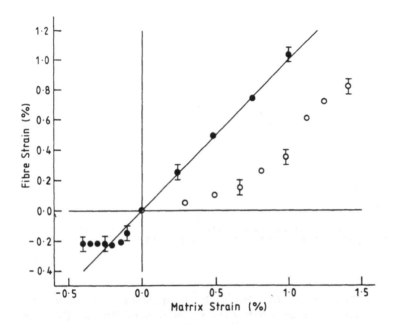

Figure 4.17 Variation of strain in the center of a fiber in a model composite with applied matrix strain; ● = room temperature–cured specimen, ○ = specimen cured at 100°C. (Data of I. M. Robinson.)

curing and cooling from the cure temperature [68]. The thermal expansion coefficients of the single crystal fibers in the chain direction are extremely small or even negative [86], and so the resin shrinkage imposes compressive stresses on the fibers parallel to their axes. The compressive stresses cause twinning in the fibers, as is shown in Fig. 16. Tensile deformation parallel to the chain direction therefore must first untwin the fibers and only then can any significant strain be taken up by the fibers [68]. This is confirmed in Fig. 16, where the twins can be seen to disappear at about 1% strain and then reappear in different places upon unloading the specimen.

The result of resin shrinkage is also demonstrated in Fig. 17, where data are given for the axial strains at the midpoints of poly-DCD fibers in an epoxy resin matrix measured using Raman spectroscopy plotted as a function of both tensile and compressive matrix strain. Two examples are given, one for a sample cured at room temperature and another for one cured at 100°C. For tensile deformation, the fiber and matrix strains are approximately equal for the room temperature–cured sample, which is not twinned. However,

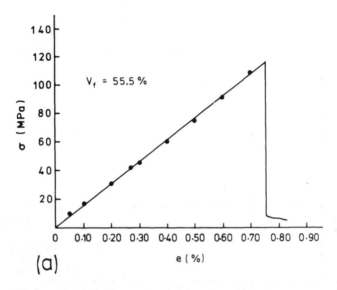

(a)

Figure 4.18 High volume fraction model composites: (a) stress-strain curve for a specimen with a 55% volume fraction of fibers, (b) composite fracture surface. (Data of I. M. Robinson.)

in the case of the sample cured at 100°C, the fiber strain falls below that in the matrix because of the presence of twins [68]. The critical strain for twin formation in poly-DCHD could be determined by loading the room temperature-cured sample in compression (Fig. 17). The fiber was first found to become compressed elastically as the Raman frequency initially increased. The fiber strain followed that of the matrix until at about 0.2% compression the strain leveled off with no further increase. At the same level of strain, twin bands were seen to appear on the fiber [68].

The Raman strain measuring technique is of interest to the study of polymer composites in general. For example, measurement of the level of tensile strain that must be applied to remove the twins gives a direct determination of the amount of shrinkage for a given resin system cured at a particular temperature. Also, since the twinning is related to the kinking found in other high modulus fibers such as aromatic polyamides in composites [87], the behavior of polydiacetylene fibers can be related directly to other important systems.

5.3 Mechanical Behavior

Measurements have been reported of the bulk mechanical properties of composites consisting of aligned poly-DCHD single crystal fibers.

20μm

(b)

A stress-strain curve for a specimen with a fiber volume fraction, V_f, of 55% is shown in Fig. 18a. It can be seen that deformation is linear up to the fracture strain of 0.8%. A scanning electron micrograph of the fracture surface of poly-DCHD/epoxy sample is shown in Fig. 18b, and it can be seen that fracture has taken place by a combination of fiber fracture and pullout. The crystalline nature of the fibers is emphasized by the faceted appearance of the fibers and holes. It is found that the mechanical properties of poly-DCHD/epoxy composites depend strongly on the fiber volume fraction; this is demonstrated in Fig. 19, where a plot is given of the Young's modulus, E, as a function of V_f. It can be seen from the figure that the stiffness of the composite increases as the volume fraction of fiber increases. The experimental points fall between the uniform strain Voigt and uniform stress Reuss lines [80]. It would be expected that for a uniaxially aligned composite sample the data should fall close to the Voigt line, and it is thought [84] that the shortfall

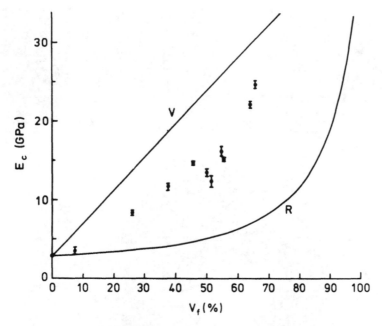

Figure 4.19 Variation of composite modulus E_C with fiber volume fraction for a poly-DCHD/epoxy aligned fiber composite. The lines V and R refer to the Voigt and Reuss averages, respectively. (Data of I. M. Robinson.)

in modulus is due to a combination of fiber misalignment and fiber twinning caused by resin shrinkage.

6. CONCLUSIONS

The high degree of molecular alignment that is found in polymer single crystal fibers leads to materials that possess high levels of stiffness and strength and are also inherently resistant to creep. It has been found that composites produced by incorporating poly-diacetylene single crystal fibers in a matrix such as epoxy resin have promising mechanical properties.

The study of these polymer single crystal fibers has led to significant advances of our understanding of the structure/property relationships in polymers and also the development of a new type of reinforcing fiber. Investigation of the structure of these materials using electron microscopy has allowed a detailed study of defects such as dislocations to be made. It has also led to the development of high resolution techniques that allow molecular detail to be seen at a level hitherto unobtainable with polymers. The analysis of surface defects such as steps has enabled estimates to be made of the theoretical strength of fibers and hence the strength of individual molecules. The frequencies of the Raman active, main chain stretching modes have been found to be a strong function of externally applied strain. This has allowed the fibers to be used to follow the micromechanics of fiber reinforcement in model composites in detail.

One of the most exciting aspects of this work is that these studies on polydiacetylene single crystals fibers have helped to open up new areas of research in more conventional polymer systems. For example, there has recently been an increase in interest in the application of high-resolution transmission electron microscopy to a wide range of polymer crystals [88]. It has also been found that the Raman strain measurement technique is applicable to other types of high modulus polymer fibers such as aromatic polyamides [89] and even carbon fibers [90]. It is clear that this will now allow detailed studies to be made of the micromechanics of reinforcement of composite systems reinforced with conventional fibers using the Raman technique.

ACKNOWLEDGMENTS

Much of the work on single crystal fibers was performed by the author at Queen Mary College. The author is grateful to D. Bloor and D. N. Batchelder for their help with supply of materials and friendly collaboration. He would also like to thank R. T. Read, C. Galiotis, P. H. J. Yeung, and I. M. Robinson for allowing him to use their results in the work described above.

REFERENCES

1. J. M. Andrews and I. M. Ward, *J. Mater. Sci.*, *5*, 411 (1970).
2. G. Capaccio and I. M. Ward, *Polym. Eng. Sci.*, *15*, 219 (1975).
3. G. Capaccio, G. Gibson, and I. M. Ward, in *Ultra-High Modulus Polymers* (A. Ciferri and I. M. Ward, Eds.), Applied Science, London, 1979, p. 1.
4. A. E. Zachariades, W. T. Mead, and R. S. Porter, in *Ultra-High Modulus Polymers* (A. Ciferri and I. M. Ward, Eds.), Applied Science, London, 1979, p. 77.
5. W. T. Mead, C. R. Desper, and R. S. Porter, *J. Polym. Sci., Polym. Phys. Ed.*, *17*, 859 (1979).

6. P. J. Lemstra and R. Kirschbaum, in *Developments in Oriented Polymers—2* (I. M. Ward, Ed.), Applied Science, London, 1987, p. 39.
7. A. J. Pennings and K. E. Meihuizen, in *Ultra-High Modulus Polymers* (A. Ciferri and I. M. Ward, Eds.), Applied Science, London, 1979, p. 117.
8. J. R. Schaefgen, T. I. Bair, J. W. Ballou, S. L. Kwolek, P. W. Morgan, M. Panar, and J. Zimmerman, in *Ultra-High Modulus Polymers* (A. Ciferri and I. M. Ward, Eds.), Applied Science, London, 1979, p. 173.
9. J. F. Wolfe, B. H. Loo, and F. E. Arnold, *Macromolecules*, *14*, 915 (1981).
10. S. R. Allen, R. J. Farris, and E. L. Thomas, *J. Materials Sci.*, *20*, 2727 (1985).
11. G. Wegner, *Pure Appl. Chem.*, *49*, 443 (1977).
12. G. Wegner, *Z. Naturforsch.*, *24b*, 84 (1969).
13. D. Bloor, L. Koski, G. C. Stevens, F. H. Preston, and D. J. Ando, *J. Mater. Sci.*, *10*, 1678 (1975).
14. D. Bloor, *Chem. Phys. Lett.*, *42*, 174 (1976).
15. K. Lochner, H. Bassler, B. Tieke, and G. Wegner, *Phys. Stat. Sol.*, *B88*, 635 (1978).
16. K. J. Donovan and E. G. Wilson, *Phil. Mag.*, *B44*, 9 (1981).
17. B. Movaghar, D. W. Murray, K. D. Donovan, and E. G. Wilson, *J. Phys. C*, *17*, 1247 (1984).
18. D. Bloor, *Phil. Trans. Roy. Soc.*, *A314*, 51 (1985).
19. E. H. Andrews and G. E. Martin, *J. Mater. Sci.*, *8*, 1315 (1973).
20. Y. R. Patell and J. M. Schultz, *J. Macromol. Sci. Phys.*, *B7*, 445 (1973).
21. H. Nakanishi, Y. Suzuki, and M. Hasgawa, *J. Polym. Sci.*, *A1*, *7*, 753 (1969).
22. A. Kelly and G. W. Groves, *Chrystallography and Crystal Defects*, Longman, London, 1970.
23. M. Dudley, J. M. Sherwood, D. Bloor, and D. Ando, *J. Mater. Sci. Lett.*, *1*, 479 (1982).
24. W. Schermann, J. O. Williams, J. M. Thomas, and G. Wegner, *J. Polym. Sci.*, *Polym. Phys. Ed.*, *13*, 753 (1975).
25. K. C. Yee and R. R. Chance, *J. Polym. Sci.*, *Polym. Phys. Ed.*, *16*, 431 (1978).
26. C. Galiotis, R. T. Read, P. H. J. Yeung, I. F. Chalmers, and D. Bloor, *J. Polym. Sci.*, *Polym. Phys. Ed.*, *22*, 1589 (1984).
27. C. Galiotis, R. J. Young, D. J. Ando, and D. Bloor, *Makromol. Chem.*, *184*, 1083 (1983).
28. C. Galiotis, R. J. Young, and D. N. Batchelder, *J. Polym. Sci.*, *Polym. Phys. Ed.*, *21*, 2483 (1983).

29. C. Galiotis and R. J. Young, *Polymer*, *24*, 1023 (1983).
30. G. C. Stevens and D. Bloor, *Chem. Phys. Lett.*, *40*, 37 (1976).
31. R. H. Baughman, *J. Chem. Phys.*, *68*, 3110 (1978).
32. G. Wenz and G. Wegner, *Makromol. Chem.*, *Rapid Commun.*, *3*, 231 (1982).
33. R. H. Baughman, H. Gleiter, and N. Sendfeld, *J. Polym. Sci.*, *Polym. Phys. Ed.*, *13*, 1871 (1975).
34. R. T. Read and R. J. Young, *J. Mater. Sci.*, *14*, 1968 (1979).
35. D. Kobelt and E. F. Paulus, *Acta Cryst.*, *B30*, 232 (1974).
36. E. Hadicke, H. C. Mez, C. H. Krauch, G. Wegner, and J. Kaiser, *Angew. Chem.*, *83*, 253 (1971).
37. P. A. Apgar and K. C. Yee, *Acta Cryst.*, *B34*, 957 (1978).
38. B. Wunderlich, *Macromolecular Physics, Vol. 1*, Academic Press, London, 1973.
39. D. T. Grubb, *J. Mater. Sci.*, *9*, 1715 (1974).
40. R. T. Read and R. J. Young, *J. Mater. Sci.*, *19*, 327 (1984).
41. R. T. Read and R. J. Young, *J. Mater. Sci.*, *16*, 2922 (1981).
42. C. J. Frye, I. M. Ward, M. G. Dobb, and D. J. Johnson, *J. Polym. Sci.*, *Polym. Phys. Ed.*, *20*, 1677 (1982).
43. M. G. Dobb, D. J. Johnson, and B. P. Saville, *Phil. Trans. Roy. Soc. Lond.*, *A294*, 483 (1980).
44. R. J. Young and P. H. Yeung, *J. Mater. Sci. Lett.*, *4*, 1327 (1985).
45. P. H. J. Yeung and R. J. Young, *Polymer*, *27*, 202 (1986).
46. A. Klug and D. J. Derosier, *Nature*, *212*, 29 (1966).
47. P. H. Lindenmeyer, *J. Polym. Sci.*, *C15*, 109 (1966).
48. V. F. Holland and P. H. Lindenmeyer, *J. Appl. Phys.*, *36*, 3049 (1965).
49. J. Petermann and H. Gleiter, *Phil. Mag.*, *25*, 813 (1972).
50. R. J. Young and J. Petermann, *J. Polym. Sci.*, *Polym. Phys. Ed.*, *20*, 961 (1982).
51. R. J. Young, R. T. Read, and J. Petermann, *Inst. Phys. Conf. Ser. No. 61*, 1981, chap. 10, p. 475.
52. R. J. Young, R. T. Read, and J. Petermann, *J. Mater. Sci.*, *16*, 1835 (1981).
53. D. N. Batchelder and D. Bloor, *J. Polym. Sci.; Polym. Phys. Ed.*, *17*, 569 (1979).
54. C. Galiotis, Ph.D. thesis, University of London, 1982.
55. A. J. Kinloch and R. J. Young, *Fracture Behavior of Polymers*, Applied Science, London, 1983.
56. F. C. Frank, *Proc. Roy. Soc.*, *A319*, 127 (1970).
57. R. J. Young, *Introduction to Polymers*, Chapman & Hall, London, 1981.
58. A. Odajima and T. Maeda, *J. Polym. Sci.*, *C15*, 55 (1966).
59. L. R. G. Treloar, *Polymer*, *1*, 95 (1960).

60. R. J. Leyrer, G. Wegner, and W. Wettling, *Ber. Bunsenges. Phys. Chem.*, *82*, 697 (1978).
61. V. Enkelmann, R. J. Leyrer, G. Schleier, and G. Wegner, *J. Mater. Sci.*, *15*, 168 (1980).
62. W. Rehwald, A. Vonlanthen, and W. Meyer, *Phys. Stat. Sol. (a)*, *75*, 219 (1983).
63. P. B. Bowden and R. J. Young, *J. Mater. Sci.*, *9*, 2034 (1974).
64. R. J. Young, D. Bloor, D. N. Batchelder, and C. L. Hubble, *J. Mater. Sci.*, *13*, 62 (1978).
65. R. J. Young, R. Dulniak, D. N. Batchelder, and D. Bloor, *J. Polym. Sci., Polym. Phys. Ed.*, *17*, 1325 (1979).
66. M. A. Wilding and I. M. Ward, *Polymer*, *19*, 969 (1978).
67. D. W. Woods, W. K. Busfield, and I. M. Ward, *Plast. Rubb. Process. Appl.*, *5*, 157 (1985).
68. I. M. Robinson, P. H. J. Yeung, C. Galiotis, R. J. Young, and D. N. Batchelder, *J. Mater. Sci.*, *21*, 3440 (1986).
69. M. G. Dobb, D. J. Johnson, and B. P. Saville, *Polymer*, *22*, 960 (1981).
70. A. Keller, personal communication.
71. M. Pietralla, *Coll. Polym. Sci.*, *254*, 249 (1976).
72. M. Bevis, *Coll. Polym. Sci.*, *256*, 234 (1978).
73. R. J. Young, in *Developments in Polymer Fracture* (E. H. Andrews, Ed.), Applied Science, London, 1979.
74. D. N. Batchelder, personal communication.
75. S. S. Brenner, *J. Appl. Phys.*, *33*, 33 (1962).
76. A. Kelly, *Strong Solids*, Clarendon Press, Oxford, 1966.
77. D. M. Marsh, in *Fracture in Solids* (D. C. Drucker and J. J. Gilman, Eds.), Interscience, New York, 1963.
78. A. J. Kinloch and R. J. Young, *Fracture Behaviour of Polymers*, Applied Science, London, 1983.
79. H. H. Kausch, *Polymer Fracture*, Springer-Verlag, Berlin, 1978.
80. D. Hull, *An Introduction to Composite Materials*, Cambridge University Press, Cambridge, 1981.
81. N. H. Ladizesky and I. M. Ward, *Pure Appl. Chem.*, *57*, 164 (1985).
82. J. H. Greenwood and P. G. Rose, *J. Mater. Sci.*, *9*, 1809 (1974).
83. C. Galiotis, P. H. J. Yeung, R. J. Young, and D. N. Batchelder, *J. Mater. Sci.*, *19*, 3640 (1984).
84. I. M. Robinson, C. Galiotis, R. J. Young, and D. N. Batchelder, *J. Mater. Sci.*, in press.
85. H. L. Cox, *Br. J. Appl. Phys.*, *3*, 72 (1952).
86. D. N. Batchelder, *J. Polym. Sci., Polym. Phys. Ed.*, *14*, 1235 (1976).
87. S. J. DeTeresa, S. R. Allen, R. J. Farris, and R. S. Porter, *J. Mater. Sci.*, *19*, 57 (1984).

88. E. L. Thomas, *Polymer Prep.*, *26*, 314 (1985).
89. C. Galiotis, I. M. Robinson, R. J. Young, B. E. J. Smith, and D. N. Batchelder, *Polym. Commun.*, *26*, 354 (1985).
90. I. M. Robinson, M. Zakikhani, R. J. Day, R. J. Young, and C. Galiotis, *J. Mater. Sci. Lett.* *6*, 1212 (1987).

5

ALUMINUM OXIDE FIBERS

JAMES C. ROMINE / E. I. du Pont de Nemours & Company, Inc.,
Wilmington, Delaware

1. INTRODUCTION

Aluminum oxide fibers constitute a very important category of fibrous
ceramic materials. The purpose of this review is to describe the
most recent trends in these fibers. Fibers based on aluminum oxide
have had a long history, and much has been written on the develop-

ment of these materials. Several comprehensive reviews detail the
earlier work [1,2]. A number of fiber forms and chemical compositions
have been included in the category labeled aluminum oxide fibers.
For the purpose of this discussion, the term fibers applies to continuous monofilaments and multifilament yarns, short fibers or staple
fibers (lengths ranging from tenths of an inch to several inches),
and whiskers (filamentary crystals with length-to-diameter ratios
greater than 10). The chemical composition of aluminum oxide-
containing fibers is quite broad, with examples ranging from fibers
of pure aluminum oxide to those that are less than half aluminum
oxide.

  Aluminum oxide has long been recognized as a superior material
in terms of chemical and thermal inertness. This inertness is particularly useful in applications where materials are exposed to oxygen
or other reactive elements, such as molten metals, at very high
temperatures. Unfortunately, processing pure aluminum oxide into
fibrous form is difficult. Since aluminum oxide is not easily processed
by most common fiber forming techniques, only a limited number
of examples of pure aluminum oxide fibers is available. Techniques
have been developed to render aluminum oxide more processable
by mixing with other oxide materials, such as silica. Consequently,
many aluminosilicate fibers are now being made, some of which use
naturally occurring aluminosilicates, such as mullite or kaolinite,
as starting materials. These aluminosilicate fibers have found application primarily as insulation and building materials and are available
at commodity prices. A description of this type of refractory fiber
can be found in several comprehensive reviews of inorganic fibers
[3,4].

  The properties of aluminum oxide fibers are significantly affected
by the addition of silica, as is illustrated by Fig. 1, which relates
fiber modulus to silica content. While this figure shows the loss
of modulus with increasing silica content, tensile strength and toughness have been found to increase with the addition of some silica.
Nonetheless, fibers with significant levels of silica are markedly
inferior in terms of refractory character and, thus, are generally
excluded from the category of high-performance aluminum oxide
fibers. This discussion will cover fibers containing at least 85%
aluminum oxide. Fibers meeting this criterion are generally those
that best take advantage of the high temperature stability and
chemical inertness of aluminum oxide.

  A special note should be made of the recent development of
alumina-silica-boria fibers. Fibers based on this composition have
shown considerable promise as composite reinforcement and specialty
insulation. These are now commercially available from the 3M company
under the trade name Nextel. These fibers contain only about 60%

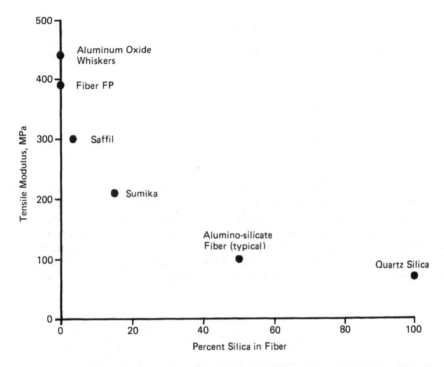

Figure 5.1   Modulus of aluminum oxide fibers as a function of silica content.

aluminum oxide and will not be discussed here. A thorough review of the topic is available elsewhere [5].

The discussion presented in this chapter concentrates on fibers that provide superior properties and value in demanding applications. Particular attention will be paid to those fibers that are being produced commercially or in significant developmental quantities. These represent the state of the art in aluminum oxide fibers and are the ones that have been the most fully described in the literature. Only four high-performance aluminum oxide fibers can be said to be in commercial or near-commercial production at this time. These are Fiber FP by du Pont, Saffil by ICI, Sumika by Sumitomo, and Denka Alcen by Denki Kagaku Kagyo. Very little is known of the Denka Alcen, as its production has only recently been announced. The review will include a description of the fiber making processes and the chemical and physical structures these processes impart to the fibers. Fiber properties will be presented in light of the requirements of the emerging applications. The discussion of applications will focus on both current uses and areas of active development. Finally, some attempt will be made to predict the future trends likely to develop in the rapidly emerging area of high-technology aluminum oxide fibers.

2. PREPARATION AND PROCESSES

Although a variety of methods have been used to produce aluminum oxide fibers, no fundamentally new processes have been developed in the past decade. The methods most frequently encountered are melt, slurry, and precursor processing, and, in the case of whiskers, crystal growth by a number of means. For aluminum oxide fibers, the method of preparation greatly affects the chemical composition and crystalline morphology of the final product. The process also determines fiber form (continuous, discontinuous, or whisker) and the economics of production. The major categories of fiber making processes will be discussed with some examples of each process.

2.1 Melt Processing

Aluminum oxide melts at 2045°C to give a low viscosity fluid. This excessive processing temperature and the poor fiber forming nature of molten aluminum oxide have restricted the use of melt processing. Some mention has been made of aluminum oxide fibers extruded [6] or drawn [7] from the melt, but these appear to be restricted to laboratory preparation. In the case of fibers drawn from the melt, continuous single crystal filaments of 50–500 μm diameters were produced at a rate of 150 mm per minute. The large filament diameter of fibers produced this way results in very poor flexibility, and the slow production rate leads to high manufacturing costs. Other workers have described melt spinning aluminum metal fibers, which are then oxidized to aluminum oxide [8], but again, this technology is not being practiced commercially.

2.2 Slurry Processing

Slurry processing is a method for fiber production that does have several interesting examples. In general, this method is based on extruding a dispersion of aluminum oxide particles or other inorganic powders through a spinneret to produce an intermediate or "green" fiber. The dispersion contains a component that acts as a vehicle to provide the rheological properties needed for spinning. The vehicle can be either an organic or an inorganic polymer and is removed or converted in a postspinning step to produce an all-ceramic fiber. Additives are frequently incorporated in the spinning dispersions to stabilize the slurry, aid in sintering, and moderate grain growth. The slurry method has several advantages over other commonly used processes. The spinning mixtures contain a substantial amount of material already in ceramic form, thus reducing the extent of chemical conversion required in the postspinning steps. Fiber shrinkage and sintering temperatures and times are reduced. The starting materials for the slurry processes are usually readily available at the required level of purity. Consequently, the slurry process has the potential of being a low cost route to aluminum oxide fibers. Further, since the slurry method is similar in concept to more traditional ceramic working processes, fiber makers can take advantage of existing experience in handling and converting of slurries into high performance ceramics.

One recently reported method for applying slurry processing technology describes spinning a dispersion of aluminum hydroxide in aqueous solutions of polyvinyl alcohol, followed by firing to 1150°C [9]. More frequently, slurry processes begin with aluminum oxide itself as the dispersed phase. ALCOA has patented a slurry method for aluminum oxide fibers based on spinning a dispersion of aluminum oxide in aqueous solutions of polyethylene oxide [10-12]. ICI claims a process that uses alumina monohydrate dispersed in aqueous aluminum acetate [13-15]. None of these processes is being practiced commercially.

Most slurry processes utilize conventional fiber forming technology, such as dry spinning or wet spinning. It is the conversion of the as-spun fiber to the ultimate ceramic form that is unique to ceramic fiber production. This conversion is done by heating the as-spun fibers, usually in stages, in ovens, or in some cases by passing the yarns through a flame. The process can be either continuous or done in a batch operation while the fiber is on bobbins. The sintering conditions can greatly affect fiber properties by determining the crystal phase of the aluminum oxide in the final product.

Du Pont is presently praticing a modified slurry process to produce its polycrystalline alpha aluminum oxide fiber, Fiber FP [16,17]. The slurry consists of alpha aluminum oxide particles dispersed in an aqueous solution of a basic aluminum chloride salt $[Al_2(OH)_5Cl]$. Under the proper conditions of acidity, the basic aluminum chloride polymerizes to produce a viscous attenuable slurry with has rheology suitable for spinning fibers. The resultant fibers are heat treated to convert the aluminum chloride to aluminum oxide with the loss of hydrogen chloride and water. The fiber at this intermediate stage consists of alpha aluminum oxide particles held together by an amorphous aluminum oxide binder. By passing the fiber briefly through a flame to raise the temperature above 1500°C, the aluminum oxide is converted completely to the alpha form and the particles are sintered to produce a dense structure. A final coating of silica is applied to the fiber surface to further improve the mechanical properties. Since basic aluminum chloride is converted to aluminum oxide, the process has the advantage of using the vehicle as an additional source of aluminum oxide. In this regard the du Pont process can be considered a combination of the slurry and precursor methods.

2.3 Precursor Processing

Precursor processing is the name given to a large variety of fiber producing methods, which are based on the concept of forming fibers from nonceramic materials that are subsequently converted into the ceramic product. The types of materials that serve as ceramic precursors include metal salts, organometallics, organic polymers, and inorganic polymers. Although these materials cover a wide range of compounds, they all have in common the ability to be transformed into useful ceramics through either chemical or thermal means. The advantages of the precursor routes to aluminum oxide fibers lie in the ease of processing nonceramic systems and the wide range of chemical compositions accessible with these methods. Often the starting materials can be obtained with higher purity than the ceramic particles required in the slurry processing. Small amounts of additional ingredients can easily be introduced into the final product by tailoring the composition of the spinning mixtures. The primary disadvantages of the slurry route include the relatively high cost of producing and purifying the starting materials and the substantial material losses associated with the conversion step.

Conversion of metal salts into oxide ceramics is one of the most widely used precursor routes to ceramic fibers. Perhaps the most frequently reported process of this type is based on basic aluminum chloride solutions [18-20]. As was mentioned in connection with the process for Fiber FP, basic aluminum chloride can be polymerized to give a viscous, extensible solution. Fibers can be formed from these solutions by extrusion into hot air using conventional dry spinning. Then the as-spun fibers must be carefully converted to aluminum oxide by heating. The conversion takes place at around 320°C, but the fibers must be heated to considerably higher temperatures in order to convert the resultant amorphous aluminum oxide to the more useful crystalline forms.

Denki Kagaju Kogyo's recently announced production of an aluminum oxide fiber utilizes the basic aluminum chloride precursor strategy [21]. The spinning solution consists of aluminum hydroxide chloride, colloidal silica, and polyvinyl alcohol. Fibers are spun by

centrifugal means, and the resulting staple product is converted
to aluminum oxide by heat treating. The fibers made in this manner
contain 80-95% aluminum oxide.

A commercial precursor process is being practiced in ICI's production of the staple aluminum oxide fiber, Saffil [22-24]. This
process calls for spinning an aqueous solution of a polymer, such
as polyethylene oxide, and an aluminum salt. The salt can be a
basic aluminum chloride or acetate. To the spinning mixture a water-
soluble polysiloxane is added, which serves as a source of silica
in the final fiber. The solution is extruded into a high-velocity
gas flow, which blows the viscous fluid into short fibers. These
fibers are heated in air to 1000°C or higher to produce a fiber
that is at least 95% aluminum oxide.

Sumitomo also practices a form of precursor process to make
the continuous aluminum oxide fiber, Sumika [25]. In this process
the spinning solution consists of an inorganic polymer dissolved
in an organic solvent. The polymer is of the polyaluminoxane type
with various alkyl substituents. A silicate ester is added to provide
a source of silica. The mixture is dry spun, and the resulting
multifilament yarn is subjected to a heat treatment step to convert
the precursors to a mixture of amorphous aluminum and silicon
oxides. A final sintering step above 1000°C produces the densified
structure of the final product.

Another promising precursor system for producing aluminum
oxide fibers involves processing aluminum oxide sols and gels. In
one example of this approach, the Universal Oil Products Company
produced a porous aluminum oxide fiber that was suitable for insulation, filtration, and catalyst support by spinning a sol [26]. The
sol is prepared by partially hydrolyzing an aqueous solution of
aluminum chloride and reacting it with hexamethylenetetramine.
This mixture is concentrated to a viscous fluid that can be dry
spun. The resulting fibers are calcined at 300-1000°C.

A different approach to ceramic fiber production is practiced
in the "relic" processes. This term refers to a version of precursor
processing where organic fibers are impregnated with the precursor
species, followed by heat treatment to remove the organic materials
and consolidate the ceramic. Aluminum oxide fibers have been made
in this manner using such host fibers as rayon [27] and polyethylene
glycol [28].

2.4 Whisker Growth

Aluminum oxide whiskers were among the earliest ceramic whiskers
to be produced, and there are hundreds of examples reported in
the literature. The practice of growing single crystal ceramic whiskers

has a long history, and much of this work is reviewed elsewhere [29]; however, an overview of the technology is appropriate for this chapter. The major categories of whisker growing methods include the vapor phase reaction, vapor-liquid-solid (VLS), and melt growth mechanisms. Each of these has many variations. Aluminum oxide whiskers have been grown by the vapor phase reaction of aluminum with oxygen-containing silicon compounds [30] and from gaseous aluminum chloride in the presence of hydrogen, carbon dioxide, carbon monoxide, and chlorine [31]. The hydrothermal method relies on the high temperature hydrolysis of such compounds as aluminum fluoride [32]. Using the VLS technique, aluminum oxide whiskers were produced by condensation and oxidation of aluminum on an aluminum oxide seed crystal [33]. Whiskers, rods, and other shapes of single crystal aluminum oxide have been produced by growth from the melt [34]. Most often the melt processing is accomplished by slow withdrawal of a seed crystal from a supercooled melt of aluminum oxide. Although the substantial effort aimed at producing aluminum oxide whiskers of superior properties has been largely successful, none of these processes is being run commercially because of the unfavorable economics of the whisker making route to aluminum oxide fibers. Those whiskers that have been for sale in the past were priced at no lower than $1000.00 per pound. Very few applications can justify such a cost.

3. CHEMICAL STRUCTURE

Aluminum oxide is a widely available and commonly used ceramic material. Consequently, a great deal has been published on every aspect of its physical and chemical nature [35,36]. Those properties of aluminum oxide that are of most interest in fibrous materials include mechanical properties, heat resistance and other thermal properties, and chemical inertness. A variety of combinations of these properties is attainable with aluminum oxide because of the complicated crystal phases and morphologies it can assume.

One of the primary distinctions in the structure of aluminum oxide in fiber form is the arrangement of the aluminum and oxygen atoms into crystalline domains. Aluminum oxide ceramic structures can be classified as either amorphous, single crystal, or polycrystalline. The amorphous form is of no importance in a discussion of high performance fibers, as it is substantially inferior in properties to the crystalline forms. The single crystal aluminum oxide denotes a complete ordering of the atoms of the material with a single orientation throughout the body of the structure. For alpha aluminum oxide the single crystal form is often referred to as sapphire. Single crystal fibers are usually those grown as whiskers. Ceramics of a single crystal have highly anisotropic properties. The polycrystalline form is the most prevalent in continuous and staple aluminum oxide fibers. In this form, atoms are ordered into crystalline domains, called grains, which are bound tightly together by intergranular forces. The grains frequently are randomly oriented and thus give the ceramic structure isotropic character.

Despite the extensive study of aluminum oxide, considerable confusion still remains concerning its chemical structure when produced under a range of conditions, such as those described in the section on fiber processes. The diagram below gives some indication of the crystal transformations that can take place in the formation of aluminum oxide-based fibers.

Amorphous

	Hydrated forms
Room	Gibbsite, Bayerite
temperature	Boehmite, Diaspore
300-500°C	Chi, eta
500-800°C	Gamma
800-1100°C	Kappa, delta, theta
>1100°c	

Alpha

The most important of these phases for high performance fibers are those that are produced at higher temperatures. The crystallographic designation of these important phases is given in the following list.

Alpha: rhombohedral (corundum)
Kappa: orthorhombic
Theta: monoclinic
Gamma and delta: tetragonal

Each of these forms of aluminum oxide has its own set of physical and chemical properties. A detailed description of them is beyond the scope of this review. Some of the important aspects of the influences of the different forms of aluminum oxide on fiber structure and fiber properties will be presented in the appropriate sections.

Some mention should be made of the influence on the chemical structure of aluminum oxide by the addition of silica. This addition is common in many of the fiber making processes that have been

described. As was mentioned earlier, the addition of silica improves
the processability of aluminum oxide. In those cases where substantial amounts of silica are added, the chemical nature of the aluminum
oxide is completely altered with a lowering of the melting point
and the formation of mixed oxide compounds, such as mullite
($3Al_2O_3 \cdot 2SiO_2$). In the case of more modest additions, the silica
acts as a sintering aid. Silica substantially lowers the temperatures
required for densification by sintering and, in doing so, tends
to keep the grain sizes to a minimum. Most of the cilica in these
systems resides at the grain boundaries, which can account for
the observed lowering of the modulus of fibers that contain a few
percent silica over those that do not. The increase in tensile strength
with incorporation of silica is probably the result of reduced grain
size and lower porosity.

4. FIBER STRUCTURE

The important macrostructural features of aluminum oxide fibers
are similar to those of most other fiber systems and include filament

Figure 5.2   Electron photomicrographs of Fiber FP (3000x).

diameter, cross-sectional shape, the number of filaments per yarn bundle, and staple length. These features are important in determining the way in which the fibers can be handled, arranged into composite preform structures, woven, braided, and processed into felts, papers, or boards. In the case of whisker fibers, the diameter and the aspect ratio (length to diameter) are among the most important physical features.

The microstructural character of an aluminum oxide fiber is essentially the same as the structure of the aluminum oxide ceramic from which it is made. Fibers do have some special characteristics that might not be found in bulk ceramics, such as orientation of particles due to alignment during spinning. Fibers also are distinguished from other ceramic structures by the relatively high surface-to-volume ratio. This tends to cause the fiber properties to be strongly influenced by the nature of the fiber surface. Each fiber making process imparts to the product a specific set of structural features. These features will be discussed for the more important aluminum oxide fibers.

Fiber FP is produced as continuous yarn of 200 filaments that are 20 μm in diameter. The fiber cross section is round, and the

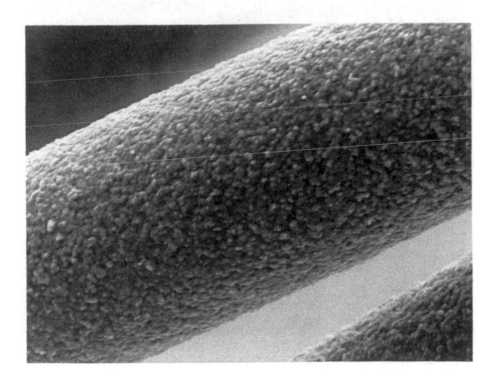

aluminum oxide phase is totally of the alpha form. Fiber FP is a
polycrystalline material with grains approximately 0.5 μm in size.
Figure 2 provides a scanning electron micrograph (SEM) of the
structure of Fiber FP. Clearly seen are the aluminum oxide grains,
which appear in a cobblestone pattern on the fiber surface. The
rough surface has been hypothesized to contribute to good adhesion
between this fiber and various matrices in composites through a
mechanical interlocking mechanism. The dominant microstructural
features of Fiber FP are the same as any dense polycrystalline
alpha aluminum oxide ceramic, and the porosity is quite low, with
its density being 98% of the theoretical density of aluminum oxide.

Saffil is a 3- m-diameter, staple fiber with filament lengths
varying from 1 to 5 cm. A distinguishing feature of Saffil is the
nearly complete absence of the nonfibrous material, referred to
as shot, which frequently accompanies fibers made by blowing or
centrifugal spinning. The prevention of shot formation in the Saffil
process has been attributed to precise control of the rheology of

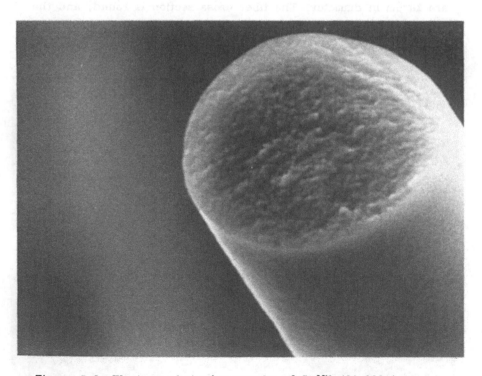

Figure 5.3   Electron photomicrographs of Saffil (20,000x).

the spinning solution and has significantly enhanced its utility as a composite reinforcement. Figure 3 shows the microstructure of Saffil, which appears nearly featureless in the SEM but which is reported to be composed of microcrystals of aluminum oxide in an amorphous matrix [26].

Sumika is another continuous, multifilament aluminum oxide yarn. The number of filaments per yarn is 380, and the fibers are of approximately 17 μm diameter. The cross section of these fibers is round, and the aluminum oxide is primarily of the gamma phase. Sumitomo claims that the process used to make Sumikau results in an intermediate fiber comprised of aluminum oxide that remains amorphous up to 900°C [37]. Upon heating above 1000°C, the aluminum oxide undergoes a transition to the gamma phase, but remains as extremely fine crystalline particles of 50 Å size. Figure 4 shows an SEM of the microstructure of Sumika, but cannot resolve any of the extremely small aluminum oxide grains that are present. The dense and highly refined structure of Sumika is thought to be responsible for its good tensile strength.

Figure 5.4 Electron photomicrographs of Sumika (3000x).

5. PHYSICAL AND CHEMICAL PROPERTIES

The physical and chemical properties of aluminum oxide-based fibers that are of interest in most applications fall into several categories, highlighted by the following list.

Physical:
  High melting point
  Low density
  Low wear/abrasive
Chemical:
  Low reactivity
  Low volatility
Mechanical:
  High strength (tensile and compressive)
  High modulus
  Low creep
  Low fatigue

Thermal:
    High use temperature
    Low thermal expansion
    High thermal conductivity
Electrical:
    Low dielectric constant
Others:
    Nonstrategic materials
    Potentially low cost

All of these properties are primarily a function of the aluminum oxide itself and can to some extent be inferred from data on bulk aluminum oxide. Of course, the combination of the aluminum oxide with other materials, such as silica, does have considerable influence on fiber properties. For the most part, extensive evaluation of fiber properties has been done only on those aluminum oxide fibers that are being produced in substantial quantities. A sizable body of property data is also available for many aluminum oxide whiskers. Table 1 provides a number of the important properties that have been reported for aluminum oxide fibers.

Table 1 Properties of Aluminum Oxide Fibers

| Property | Currently available fibers | | | |
	Fiber FP	Saffil	Sumika	Whiskers
Supplier	du Pont	ICI	Sumitomo	
Composition				
Al_2O_4 (%)	99.5	96-97	85	100
SiO_2 (%)		3-4	15	
Crystal form	alpha	delta	gamma	alpha
Fiber form	Contin. yarn	Staple	Contin. yarn	Single crystal
Filaments per yarn	200		380	
Fiber diameter (10^{-6} M)	20	3	17	0.5-30
Density (g/cm^3)	3.9	3.3	3.2	3.97
Tensile strength				
(MPa)	1,560	2,000	1,775	700-14,000
(KPSI)	220	290	250	100-2,000
Max. strain (%)	0.4	0.7	0.8	
Tensile modulus				
(GPa)	390	300	210	440
(MPSI)	55	43	30	62
Compressive strength[a]				
(MPa)	2,500		1,420	
(KPSI)	350		200	
Dielectric constant[a] @ 10^{10} Hz	5.3		4.2	
Coef. of thermal expand. (10^{-6} °C^{-1})	6.8		8.8	
Thermal conductivity (kcal/m hr °C)	26.7			
Refractive index, n_D^{20}	1.7		1.65	

[a]These properties were measured on unidirectional epoxy resin composites with 35 vol% fiber.

An examination of the physical properties of available aluminum oxide fibers reveals that they have exceptionally high tensile modulus and compressive strength, but have only moderate tensile strength and low strain to failure. Other high-performance reinforcing fibers, such as carbon and silicon carbide, may provide a somewhat better combination of properties for certain composite applications. The advantages that aluminum oxide fibers hold over these potential competitors lie in thermal and chemical stability. The upper use temperatures for aluminum oxide fibers in an oxidizing atmosphere are claimed to be 1000-1600°C. Neither carbon nor silicon carbide fibers is able to tolerate such exposures. This type of high temperature stability is of greatest importance in two emerging application, specialty insulations and ceramic composite reinforcement. In contact with molten metals, such as aluminum, magnesium, iron, and titanium, aluminum oxide fibers show the greatest resistance to chemical attack of all of the available ceramic fibers. When used as reinforcement in metal matrix composites, aluminum oxide fibers do not require special coatings or fabrication procedures to avoid property degradation. Aluminum oxide fibers have even been used as the filtration medium for some metal refining operations. Some degradation is seen when aluminum oxide fibers come into contact with molten alkali metals, and aluminum oxide is known to be attacked by hot concentrated mineral acids and caustic solutions. Considering the overall combination of good physical properties and thermal and chemical stability, aluminum oxide fibers are of keen interest to designers and engineers.

6. APPLICATIONS AND MARKETS

Applying the unique set of properties available in aluminum oxide fibers to demanding engineering problems is an area of significant activity. Thus far, the majority of applications have been as reinforcement in composite structures; however, some noncomposite use, such as specialty insulations and catalyst supports, are also important.

6.1 Composites

A continuing thrust in material science has been the development of composites that can overcome the deficiencies of more traditional engineering materials. Now the drive is toward lighter, stronger, and stiffer materials for use in applications at elevated temperatures [38]. The fatigue performance, toughness, damage tolerance, and dimensional stability that the composites offer are of great practical importance [39]. Aluminum oxide fibers provide the potential for substantial improvements in composite properties for many applications [40]. The incentives for development of new materials containing aluminum oxide fibers include weight reduction [41], simplification of component fabrication [42], and improvements in the performance of entire systems. Performance is of particular concern when dealing with high temperature applications [43]. Reliability is another key factor in many designs, especially those of critical structural components and systems that cannot be serviced easily, such as space structures [44]. Much has already been written about the use of ceramic fibers in composite systems [45-47]. These systems are generally classified by the type of matrix that is reinforced: metal, resin, or ceramic. Each of these classes of composites has its own unique set of properties, applications, and problems. The following sections discuss the use of aluminum oxide fibers in reinforcement of these matrix systems.

6.1.1 Metal Matrix Composites

Reinforcement of metals with aluminum oxide fibers has been an area of active development since the mid-1960s, yet commercial adoption of metal matrix composites (MMC) has been relatively slow. The problems have been both technical and economic, with the development of fabrication technology and the high cost of fibers being two major impediments to penetration of the structural metals market. The advantages of MMCs over unreinforced metals are numerous and include improved strength and stiffness, better high temperature properties, much improved fatigue life, creep resistance, tailored thermal properties, and better wear performance.

Aluminum oxide fibers are important candidates for reinforcement of metals because of their inertness to oxygen and molten metals at elevated temperatures. These fibers also provide superior stiffness because of their high modulus. They have been evaluated a number of metals such as aluminum, magnesium, copper, and lead. The fabrication of composites from aluminum oxide fibers and these metals has been the subject of a great deal of work in the past two decades [48,49].

At present, the list of fabrication methods that have been evaluated for MMCs include solid phase techniques such as diffusion bonding, hot rolling, extrusion, drawing, hot isostatic pressing, and explosive welding; and liquid phase techniques including vacuum or pressure infiltration, squeeze casting, and rheocasting or compocasting [50]. Much of the development work in MMCs has come out of programs sponsored by the U.S. government. Most of these relate to military and space applications. One such program being conducted by the Boeing Vertol Company is aimed at reduction of noise and vibration in helicopter transmission housings [51]. In

this program, full-size engine transmission housings were made
of aluminum oxide fiber-reinforced magnesium (Fiber FP/Mg). These
16-in.-diameter components represent the largest continuous fiber-
reinforced MMC ever made by casting. The housings are still undergoing evaluation, but the advantage of the Fiber FP/Mg system
is clear. The MMC material provides nearly three times the specific
stiffness of the unreinforced metal and can maintain its stiffness
to much higher temperatures. Jet engine turbine blades have been
an area of considerable MMC work. The goal of a number of
government-sponsored programs has been to produce lighter, stiffer,
and higher temperature blades to increase the fuel efficiency and
performance of jet engines. Both continuous fiber and whisker
reinforcement have been evaluated. While systems of ceramic fibers
in metals have been successfully made, these have not been adopted
because of the poor damage tolerance and brittle failure of these
materials. The tendency of MMCs to fail catastrophically is the result
of brittleness of the aluminum oxide fibers. Short fiber and whisker
reinforced systems are less sensitive to this problem, but it continues
to present a serious limitation to all MMC materials.

Automobile engines have also been targeted at applications where
aluminum oxide-reinforced metals can add value and performance.
In 1982 Toyota Motor Company began the first mass production
of an MMC component by introducing a line of diesel engines utilizing
MMC pistons. The pistons have the top land and top ring groove
areas reinforced with either aluminum oxide (Saffil) or aluminosilicate
staple fibers [52]. The purpose of this reinforcement is to improve
wear and seizure resistance, and to provide high temperature strength
and dimensional stability in the critical areas in the piston. Another
recently reported application under development is fiber-reinforced
aluminum connecting rods [53]. For this use, weight reduction is
the primary motivation. Connecting rods that are comparable in
properties to incumbent steel rods can be produced from Fiber FP-
reinforced aluminum at one-half the weight. The reduction in weight
of a reciprocating engine component can dramatically improve fuel
economy and performance.

Other MMC applications for aluminum oxide fibers have been
investigated. These include reinforcing lead for better creep resistance in batteries [54] and various metals for machine tooling.
Sporting equipment is yet another area where high performance
MMCs have been evaluated. Although commercial adoption of MMCs
has been slow, the technology has now demonstrated considerable
utility and will likely expand in the near future.

6.1.2 Resin Matrix Composites

At present, the most highly developed composite technology is that
based on reinforced resin matrices. These matrix systems are commonly

either epoxies or thermoplastics. Applications utilizing asbestos and glass fiber-reinforced resins are now mature businesses. The aerospace industry has led in the development of resin matrix composites that employ high performance, organic and carbon fibers as reinforcement. These material provide the industry with significant value in many weight saving, structural uses. Despite this considerable development, few systems based on ceramic fiber-reinforced resins have been adopted. In resins, ceramic fibers, such as those of aluminum oxide, offer few advantages over less costly alternatives. Since the resin composites are neither fabricated nor used at elevated temperatures, the thermal stability of the fibers is not a significant factor. For most stiffness-related applications, carbon fibers offer adequate properties in resins at substantially lower density. In strength and toughness critical uses, the aramid organic fibers, such as Kevlar,* are generally selected. Even though substantial penetration of the resin matrix composite market is not likely for high-performance aluminum oxide fibers, in some very specialized applications this type of reinforcement may have considerable value. One such application is in electronic circuit boards, where low thermal expansion and good thermal conductivity are important. Resins reinforced with aluminum oxide and other ceramic fibers have been considered here over carbon fibers because of the electrical conductivity of carbon. These fibers also have been evaluated in high-performance sporting equipment, such as tennis rackets and ski poles. An important new approach that utilizes the strengths of both organic fibers and inorganic fibers in resin matrixes are the hybrid composites, such as those made of Kevlar/Fiber FP [55,56].

6.1.3 Ceramic Matrix Composites

An intensive effort is currently underway to develop new ceramic materials for application to engineering problems. Japan has a national policy that emphasizes commercialization of advanced ceramics. Much of the drive in the United States comes from the U.S. government, which has targeted ceramics as the next generation of high technology materials to be used in aerospace, weapon systems, and high-efficiency automobile engines. While ceramics do offer considerable advantage in high-temperature structural uses, most of today's ceramics suffer from the severe limitations of low toughness and poor damage tolerance [57]. The prevailing approach to improving these deficiencies is by toughening the materials through fiber reinforcement [58]. Not many ceramic matrix composites (CMCs)

---

*du Pont registered trademark.

have actually been made and tested in applications, but a number of important programs are presently underway. A good assessment of the future directions of CMCs was published by the Department of Defense in 1977 [59]. The most advanced CMC systems are those of ceramic fiber-reinforced glasses [60]. Aluminum oxide fibers were evaluated in glass matrix composites, but nonoxide fibers, such as silicon carbide or silicon nitride, appear to be the leading candidates at this time. Aluminum oxide whiskers were extensively evaluated in programs aimed at developing CMC gas engine turbine blades. Improved high temperature properties were demonstrated with these materials, but the composites were prone to catastrophic failure. While ceramic composites are certainly in an embryonic stage, the thrust toward much higher temperature applications will most certainly advance the technology. Since few materials can match the thermal and chemical stability of aluminum oxide, fibers of this type are likely to play an important role as reinforcement in ceramic matrices.

6.2 Noncomposites

High-performance aluminum oxide fibers have found use in applications other than composite structures. The majority of these non-composite applications are for specialty insulations. The less expensive aluminosilicate fibers dominate the high-performance insulation market, but under certain circumstances the thermal stability and inertness of aluminum oxide is required [61,62]. Much of the staple aluminum oxide fiber produced is used for high-temperature furnace insulation in the metal and ceramic industries [63]. Saffil is frequently used in a variety of forms as specialty insulation. Aluminum oxide fiber products for insulation are sold in cloth, yarn, felt, paper, and board form.

Additional noncomposite applications for aluminum oxide fibers include asbestos replacement as insulation and as friction materials in brakes and clutches. Aluminum oxide fibers have been used for catalyst supports where the inert and thermally stable surface of these fibers is well suited for catalyst attachment. With the high surface-to-volume raio of fibers, these catalyst support structures can be made highly efficient. The same principle applies to aluminum oxide fiber mantles for flameless heating systems.

7. FUTURE TRENDS

The future looks bright for fibers made of aluminum oxide and related ceramic materials. Several factors contribute to this conclusion.

Aluminum oxide has a well-recognized combination of superior properties, especially for high temperature applications, and current trends
in materials development are toward much higher use temperatures.
Aluminum oxide is an abundant, nonstrategic material available at
commodity prices that, coupled with the proper processing technology,
should afford relatively inexpensive ceramic fibers. Several aluminum
oxide fibers are now being produced commercially, a fact that should
hasten the development of many applications. The future trends
for development of new aluminum oxide fibers will include improvements in strength and toughness, especially at elevated temperatures.
These improvements will come through the application of techniques
such as control of microstructure, transformation toughening, and
fiber surface modification. Lower cost processes can be expected
as the economy of scale is achieved through increased production,
and totally new routes to aluminum oxide fibers may even be developed as a result of the substantial increase in emphasis on ceramic
research. As existing materials are stretched to the limits of performance, aluminum oxide fibers appear well suited to contribute
to the high technology solution of difficult engineering problems.

REFERENCES

1. A. G. Cottrell and B. E. Lee, *Repts. Prog. Appl. Chem.*,
   60, 50 (1975).
2. D. V. Badami, *Shirley Inst. Publ.*, *536*, 217 (1979).
3. W. C. Miller, Refractory Fibers, in *Encyclopedia of Chemical
   Technology*, Vol. 20, Wiley-Interscience, New York, 1979,
   p. 65.
4. H. F. Arledter, Inorganic Fibers, in *Encyclopedia of Polymer
   Science and Technology*, 3rd ed., Vol. 6, Wiley-Interscience,
   New York, 1967, p. 610.
5. D. D. Johnson, *J. Coated Fabrics*, *11*, 282 (1981).
6. G. Sporleder, *Haus Tech.*, *Essen. Vortragsveroeff*, *298*, 7
   (1974).
7. Tyco Laboratories, Japanese Patent 46 16 082, (1946).
8. Toyobo Company, Ltd., Japanese Patent 81 49,020, 1981.
9. Mitsubishi Light Metal Industries, Ltd., Japanese Patent
   82 47,919, 1982.
10. Aluminum Company of America, U.S. Patent 3,705,223, 1972.
11. Aluminum Company of America, Netherlands Patent 73 09, 379, 1974.
12. Aluminum Company of America, German Patent 2,225,741, 1974.
13. Imperial Chemical Industries, Ltd., German Patent 2,052,725, 1971.
14. Imperial Chemical Industries, Ltd., German Patent 2,029,731, 1971.
15. Imperial Chemical Industries, Ltd., German Patent 2,119,886, 1971.

16. A. K. Dhingra, *Phil. Trans.*, *Roy. Soc. London*, *A294*, 411 (1980).
17. E. I. du Pont de Nemours & Co., British Patent 1,264,973 (1972).
18. Toshiba Monofrax Company, Ltd., Japanese Patent 74 98,804, 1974.
19. Carborundum Company, DE 2,732,290, 1978.
20. United Aircraft Corporation, U.S. Patent 3,865,917, 1975.
21. Denki Kagaku Kogyo K. K., U.S. Patent 4,348,341, 1982.
22. Imperial Chemical Industries, Ltd., British Patent 1,360,197, 1976.
23. Imperial Chemical Industries, Ltd., U.S. Patent 3,992,498, 1976.
24. J. D. Birchall, *Proc. Br. Ceramic Soc.*, *33*, 51 (1981).
25. Sumitomo Chemical Company, Ltd., Japanese Patents 83 98,428 and 58 98,428, 1983.
26. Universal Oil Products Company, U.S. Patent 3,632,709, 1972.
27. Toray Industries, Inc., Chemical Abstracts:82(16)1023875.
28. Bayer, A. G., German Patent 2,163,678 (1973).
29. A. P. Levitt, Ed., *Whisker Technology*, Wiley-Interscience, New York, 1970.
30. S. S. Brenner, *J. Appl. Phys.*, *33*, 33 (1962).
31. W. B. Campbell, *Chem. Eng. Progr.*, *62*, 68 (1966).
32. I. Yamai and H. Saito, *J. Cryst. Growth*, *45*, 511 (1978).
33. D. J. Barber, *Phil. Mag.*, *10*, 75 (1964).
34. H. E. LaBelle, Jr., and A. I. Mlavsky, *Nature*, *216*, 574 (1967).
35. W. H. Gitzen, *Alumina as a Ceramic Material*, The American Ceramic Society, Columbus, Ohio, 1970.
36. C. T. Lynch, Ed., *Handbook of Material Science*, Vol. 2, CRC Press, Boca Raton, Fla., 1974.
37. Y. Abe, K. Fujimura, and S. Horikira, *J. Jap. Soc. Composite Mater.*, *6*, 89 (1980).
38. K. K. Chawla, *J. Metals*, *35*, 82 (1983).
39. T. W. Chou, *Mater. Sci. and Eng.*, *25*, 35 (1976).
40. A. K. Dhingra, Advances in Inorganic Fiber Developments, in *Contemporary Topics in Polymer Science*, Vol. 5 (E. J. Vandenberg, Ed.), Plenum Press, New York, 1984, p. 227.
41. G. Lubin and S. J. Dastin, Aerospace Applications of Composites, in *Handbook of Composites* (G. Lubin, Ed.), Van Nostrand Reinhold, New York, 1982, p. 772.
42. M. Martin and J. F. Dockum, Jr., Composites in Land Transportation, in *Handbook of Composites* (G. Lubin, Ed.), Van Nostrand Reinhold, New York, 1982, p. 679.
43. E. J. Kubel, Jr., *Mater. Eng.*, *100*, 47 (1984).
44. J. Persh, *Ceramic Bull.*, *64*, 555 (1985).

45. G. Lubin, Ed., *Handbook of Composites*, Van Nostrand Reinhold, New York, 1982.
46. A. A. Watts, Ed., *Commercial Opportunities for Advanced Composites*, American Society of Testing and Materials, Philadelphia, 1980.
47. L. J. Broutman and R. H. Krock, *Modern Composite Materials*, Addison-Wesley, Reading, Mass., 1967.
48. A. Kelly, *Sci. Am.*, *212*, 28 (1965).
49. T. W. Chou, A. Kelly, and A. Okura, *Composites*, *16*, 187 (1985).
50. T. W. Chou, A. Kelly, and A. Okura, *Composites*, *16*, 187 (1985).
51. A. P. Levitt, Army Metal Matrix Composite Program Overview, in *Proceedings of the Sixth Metal Matrix Composite Technology Conference, Santa Barbara, California*, 1985, pp. 1.2-1.26.
52. T. Donomoto, K. Funatani, N. Miura, and N. Miyake, Ceramic Fibre Reinforced Piston for High Performance Diesel Engines, SAE Int. Congress and Exposition, Detroit, Michigan Paper 830252, 1983.
53. F. Folgar, W. H. Krueger, and J. G. Goree, *Metal Matrix, Carbon, and Ceramic Matrix Composites*, NASA Conference Pub. 2357, 1984, p. 43.
54. H. S. Hartmann and R. A. Sutula, *J. Electrochem. Soc.*, *129*, 1749 (1982).
55. E. I. du Pont de Nemours & Co., U.S. Patent 3,556,922, 1971.
56. E. I. du Pont de Nemours & Co., U.S. Patent 3,778,334, 1973.
57. I. W. Donald and P. W. McMillan, *J. Mater. Sci.*, *11*, 949 (1976).
58. R. W. Rice, *Ceramic Bull.*, *63*, 256 (1984).
59. J. E. Hove and H. M. Davis, Assessment of Ceramic-Matrix Composite Technology and Potential DoD Application, Contract No. DAHC15-73-C0200, IDA Paper P-1307, 1977.
60. G. K. Layden and K. M. Prewo, Development of Broadband Radome Material, Final Report, AFWAL-TR-82-4100, 1982.
61. E. Rastetter and W. R. Symes, *Inter. Ceram.*, *31*, 215 (1982).
62. Nippon Asbestos Company, Ltd., Japanese Patent 79 82,725, 1972.
63. S. Howard, *Ceramic Bull.*, *62*, 623 (1983).

6

LEAD FIBERS

YOSHIKAZU KIKUCHI* and HISATAKA SHOJI[†] / Toray Industries,
Inc., Otsu-shi, Shiga, Japan

1. INTRODUCTION

Lead fibers have recently been developed by an application of a
melt spinning technique. Although melt spinning is one of the most
basic techniques for filament production, it is generally considered

Present affiliations:

*Fuji Xerox Co., Ltd., Minamiashigara-shi, Kanagawa, Japan
[†]Toray Industries, Inc., Kamakura-shi, Kanagawa, Japan

to be difficult to apply for metal filament production. Metals generally
exhibit low viscosity and high surface energy in their molten state.
Therefore, molten metal threads tend to break up into droplets.
Thus, there have been only a few commercial products manufactured
by melt spinning of metals.

However, lead exhibits some advantageous properties. Its low
melting point would offer a wide selection of the spinneret materials.
Also, its well-recognized susceptibility to air oxidation would offer
a great advantage in that the rapid formation of a thin layer film
of lead oxide resulting from air oxidation of the metal on the surface
of the molten thread prevents droplet formation. In our laboratory,
two types of lead fibers have so far been developed: one is short
fibers that can be used for sound insulation or radiation shielding
and the other is conjugated filaments comprising a polymer sheath
and a lead core with which fishing nets have been developed. In
this chapter, lead fibers will be discussed in terms of their preparation methods, structure, and applications.

2. PREPARATION OF LEAD FIBERS

2.1 Lead Metal Fibers

2.1.1 Stability of Liquid Metal Jet

A number of techniques have been proposed for melt spinning of
metals to obtain metal fibers. Such techniques are essentially of
four types: (a) free-flight melt spinning [1-17], (b) melt extraction
[10,18-21], (c) chill block melt spinning [22], and (d) the Taylor
method (glass-coated melt spinning) [23-36]. The free-flight melt
spinning technique is characterized by its simplicity and is considered
to be most promising in terms of metal fibers production.

Based on the calculation about the jet stability, Butler et al.
[8] reported that the lower limit of fiber diameter attained by this
technique would be 100 μm. On the other hand, Nagano [7] proposed
the following equation for calculation of the breaking length (Z^*):

$$Z^* = \left(\frac{2d^3P}{\alpha} \right)^{1/2} \ln\left(\frac{R}{\delta_0} \right)$$

where α, P, and d are surface tension, pressure, and diameter
of orifice, respectively. Using the value of 19.5 reported by Grant
and Middleman [37] as the $\ln(R/\delta_0)$ value, Z^* values for several
metals were calculated for the orifices with diameters of 0.005 and
0.01 cm (Table 1). Based on the calculated Z^* values, the lower
limit of orifice diameter would be 0.005 cm when a turbulent flow
is grown.

Table 6.1   Breaking Length by Jet Flow (pressure $P = 2$ kg/cm^2)

Kind of metal	Surface tension α (dyne/cm^2)	Density (g/cm^3)	Z^* (mm) $d_0 = 0.01$ cm	Z^* (mm) $d_0 = 0.005$ cm
Pb	440	10.9	18.3	6.40
Zn	770	6.7	13.9	4.85
Al	500	2.3	17.3	6.05
Cu	1103	9.0	11.5	4.02
Fe	1360	7.9	10.5	3.68

From Ref. 7.

Cunningham et al. [16] pointed out the importance of the stability of the molten metal jet in the process of filament production from inviscid metals. Lead is susceptible to air oxidation and readily oxidizes to form a thin layer film of lead oxide on the surface. The resulting lead oxide film is considered to have an effect of enhancing the stability of the jet. In fact, as seen in Fig. 1, filaments of molten lead can be obtained when the jet is extruded into an oxygen-containing atmosphere, but not into a nitrogen gas atmosphere. However, scanning electron micrographic examination of lead fibers thus obtained reveals the pearl necklace-like appearance, indicating that the lead oxide film formation is not enough to prevent the growth of varicose oscillation (Fig. 2a).

Solidifying rate is also an important factor for metal jet stabilization and, thus, for metal filament production. If the solidifying rate is fast enough, varicose oscillation would not occur. Therefore, varicose oscillation is expected to be inhibited by alloying with a metal that can increase the solidifying rate of lead. When a minute amount (140 ppm) of tin is added to lead melt, the surface of the resulting fiber exhibits no apparent growth of the varicose oscillation (Fig. 2(b). This was first interpreted to mean that the additive accelerated the solidification of lead and, thus, stabilized the jet. However, DSC analysis reveals no apparent dependence of latent heat, specific heat, or melting point on the concentration of tin (Table 2). Therefore, a mechanism other than the jet stabilization by acceleration of the solidifying rate of lead melt by the addition of tin is operating. Cunningham et al. [16] have reported that steel filament can be produced by alloying with a small amount of aluminum or silicon and that the jet is stabilized by the formation of a thin film of the alloying component. The mechanism is detailed in Sec. 3.

Vo=7.9m/sec

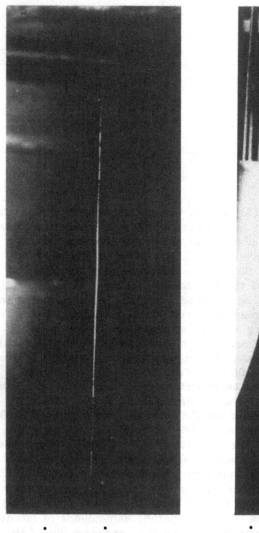

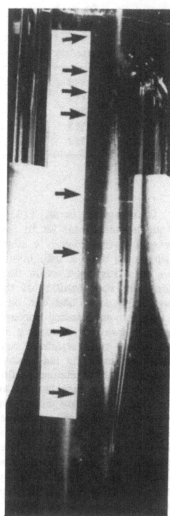

in air   in N₂gas

Figure 6.1 Stroboscopic photograph of molten Pb jet from the orifice into the air or N$_2$ gas atmosphere.

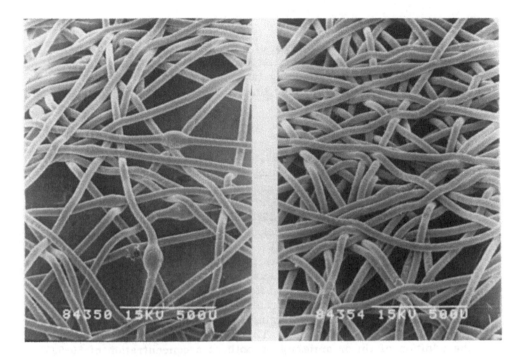

Figure 6.2 Scanning electron micrographs of lead fibers: (a) fibers from the pure lead (Pb), (b) fibers from tin-containing molten lead [Pb/Sn (140 ppm)].

Table 6.2 Thermal Properties of Pb Metal

Pb/Sn (ppm)	T_m (°C)	Latent heat (cal/g)	°C	Specific heat	
				Pb	Pb/Sn (140 ppm)
				(cal/g °C)	
0	327.8	5.48	350	0.0355	0.0353
20	327.6	5.47	354	0.0355	0.0352
70	327.6	5.46	362	0.0352	0.0351
140	327.7	5.48	370	0.0353	0.0351
			378	0.0353	0.0350

Apparatus: Perkin-Elmer DSC-2; heating rate, 10.0 °C/min; atmosphere, dry N_2 flow.

2.1.2 Spinning Apparatus

Figure 3 is a schematic diagram of a typical spinning apparatus for lead fibers. The apparatus is composed of essentially three units, a melting tank, a holding tank, and a spinning block. All units and their connecting pipe lines are equipped with a resistance heater.

2.1.3 Melt Spinning Procedures

Lead ingot with a purity over 99.99% is melted at 340°C in air and then tin metal and/or lead-tin alloy is added. Air is purged from the holding tank with nitrogen gas, to which lead melts are transported from the melting tank. The temperature of lead melts is maintained at approximately 10-20°C above the melting point. The temperature of the spinning block is adjusted so that the orifice temperature is kept at 340-350°C.

Streaming through the orifice is initiated by pressurizing the holding tank with an inert gas such as nitrogen or argon. The pressure should be adjusted to afford a reasonable jet stability as well as a desired mass flow rate. When pure lead alone is used for spinning, spontaneous plugging at the nozzle causes a rapid decrease in the mass flow rate. This problem can be solved by the addition of tin or antimony or both at a concentration of 50-500 ppm [38].

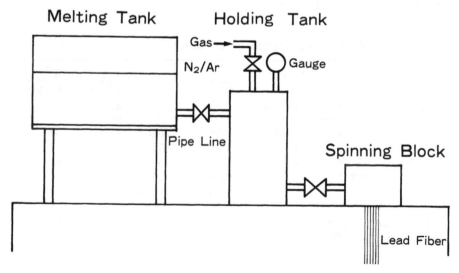

Figure 6.3 Schematic diagram of the spinning apparatus for the lead fibers.

The fiber diameter definitely depends on the orifice diameter used for spinning. Thus, spinning with 0.05- and 0.18-mm orifices yields fibers with a diameter of 40 μm and of 160 μm, respectively. Using a 0.05-mm orifice, filaments with a length of over several meters can be obtained.

2.2 Polymer Sheath, Lead Core, Conjugated Filaments

Filament having a lead continuous core surrounded by a synthetic polymer sheath can be produced by simultaneously extruding a molten lead and a molten synthetic polymer through a spinneret [39-41] (Fig. 4). The sheath synthetic polymer should have a reasonably high spinnability to compensate for the poor spinnability of lead. The core metal should melt at or below the temperature at which the sheath polymer can be extruded. The melting point of lead, 327°C, is slightly too high to extrude the sheath polymer. Therefore, various lead alloys [42-44] have been proposed as the core metal (Table 3). The molten-sheath core filament can solidify upon contact with air or water. More rapid solidification is preferable in order to prevent the formation of droplets of lead in the core due to the low viscosity and large surface tension of the molten lead.

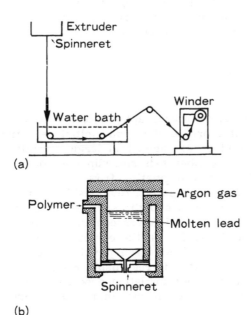

(a)

(b)

Figure 6.4 Spinning apparatus for the conjugated fibers: (a) diagram of spinning apparatus, (b) spinning block.

Table 6.3   Lead Alloys

Composition	Melting point (°C)
Pb–8%Sb	285
Pb–10%Sb	298
Pb–4%Sb–4%Sn	288
Pb–2.25%Mg	248

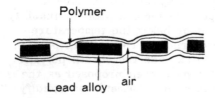

Figure 6.5   Cross-sectional view of a filament with fractured cores.

The filaments thus formed can be further subjected to a drawing process to obtain two types of filaments, filaments containing a continuous core and those containing fractured cores (Fig. 5).

3.   STRUCTURE OF LEAD FIBERS

3.1   Surface Observations

Figures 6 and 7 are scanning electron micrographs of lead fibers streamed from an 0.05-mm-diameter orifice at various initial velocities. The surface of pure lead fibers (Fig. 6) shows that large crystals are arranged in the direction of the fiber axis and slip lines of crystals are inclined due to the stress originating from the gravitational and aerodynamic forces. But no apparent relationships can be observed between the initial velocities and the surface appearance of the resulting fibers. In the case of lead fibers containing tin at a concentration of 40–140 ppm (Fig. 7), on the contrary, such slip lines are not so clearly observed. The reason for this change of surface appearance is not clear, but it is reasonably assumed that the crystalline structure of the fiber formation process is different between the pure lead fibers and the tin-containing lead fibers.

3.2   X-Ray Photoelectron Spectra (XPS) of Lead Fibers

X-ray photoelectron spectroscopy (XPS) is one useful technique to obtain chemical information about the surface layers of lead fibers.

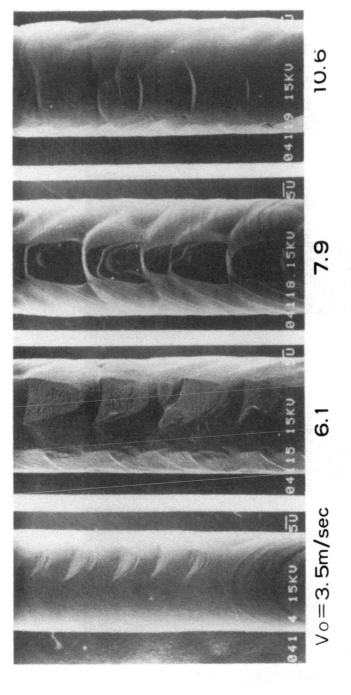

Figure 6.6 Scanning electron micrographs of the surfaces of pure lead fibers streamed at various initial jet velocities.

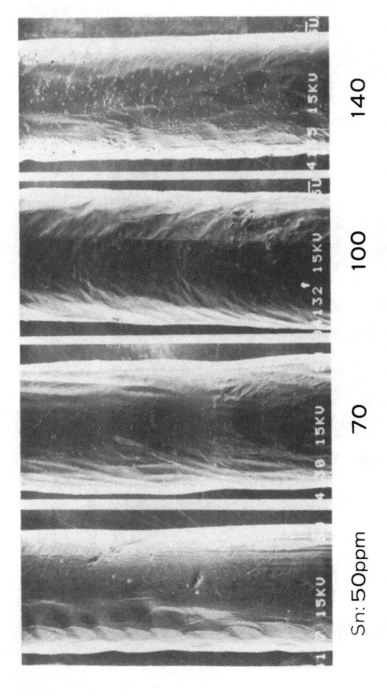

Figure 6.7 Scanning electron micrographs of changes of the surface appearances depending on the tin content in the tin-containing lead fibers.

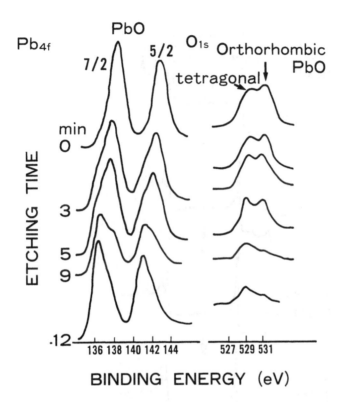

Figure 6.8   XPS spectra of $Pb_{4f}5/2,3/2$ and O_{1s}. X-ray: AlK_0; sample current of the Ar^+ etching condition: 6 μA; etching rate: about 30 Å/min.

Figure 8 illustrates XPS spectral changes of a pure lead fiber with time upon etching of the surface of the fiber with Ar^+ ion. The peak intensity ratio of O_{1s} to Pb_{4f} and the Pb_{4f} binding energy shift clearly indicate that the surface is covered with lead oxide layers [45]. The spectral changes in the region of Pb_{4f} reveal a rapid appearance of a peak attributed to lead metal as a shoulder peak of lead oxide at a slightly low binding energy region. The increase in intensity of the peak attributed to lead metal is accompanied by a decrease in intensity of peaks due to lead oxide. The thickness of the lead oxide layer is estimated to be about 300 Å from the etching rate of 30 Å/min.

As seen in Fig. 8, the O_{1s} region of the spectrum exhibits two peaks, indicating the presence of two distinct crystal forms in the PbO layer. From the binding energy of the O_{1s} electron, the higher

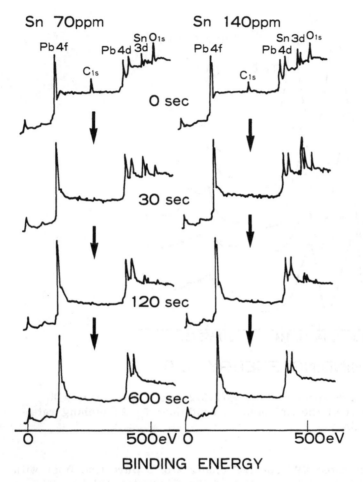

Figure 6.9 Wide-range XPS spectra of Pb/Sn (140 ppm) fibers. Sample current in Ar⁺ ion etching: 6 μA.

energy peak is attributed to the orthorhombic form and the lower to the tetragonal [45]. Spectral changes of O_{1s} show that the tetragonal form predominates in the inner layer of the PbO film and the orthorhombic in the outer layer. Therefore, the tetragonal form of lead oxide in the outer layer seems to be changed to the orthorhombic form by exposure to oxygen in the air.

Figure 9 shows spectral changes upon Ar⁺ ion etching of lead fibers containing tin metal of 70 and 140 ppm, respectively. In both cases, the intensity of the Sn_{3d} peak decreases with etching time, and finally the Sn_{3d} peak almost entirely disappears at 600 sec etching. From the Sn_{3d} spectrum shown in Fig. 10, the chemical composition of tin element was confirmed to be SnO_2.

In Fig. 11 the Pb_{4f} spectra were also recorded and compared to the spectra of pure lead fibers. The depth profile of the Sn_{3d} and O_{1s} spectra shown in Fig. 12, together with the above observation (Figs. 9-11), give rise to the following information about the surface structure of tin-containing lead fibers.

1. Tin atoms are segregated in the outer layer of the fiber and form the SnO_2 layer of 200-300 Å thickness beneath the outermost layer.

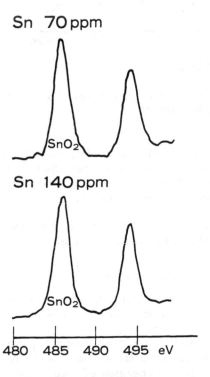

Sn 70 ppm

SnO_2

Sn 140 ppm

SnO_2

480   485   490   495 eV

BINDING ENERGY

Figure 6.10   XPS spectra of $Sn_{3d}5/2,3/2$. Binding energy of Sn_{3d}: 485.8 eV.

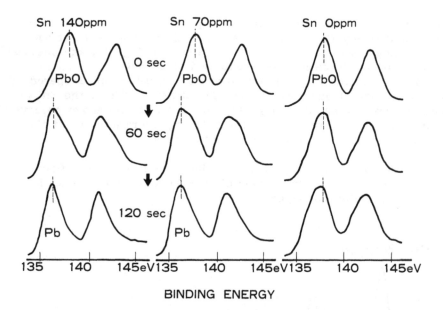

Figure 6.11 XPS spectra of Pb_{4f} 5/2,7/2 in Pb, Pb/Sn (70 ppm) and Pb/Sn (140 ppm) fibers. Sample current: 6 μA.

2. The thickness of PbO film in the tin-containing lead fibers is about tens of angstroms and is much thinner than that of the pure lead fibers. By the addition of tin metal, the oxidation of lead is thought to be interrupted.

Figure 13 illustrates the model of surface structural changes in the section of fibers based on the above consideration.

3.3 Segregation of Tin Metal and Jet Stabilization

Tin was demonstrated to be localized in the outer layer of the tin-containing lead fibers in the previous section. Here we discuss the mechanism of segregation of tin element in the fiber.

As illustrated in Fig. 14, when the molten lead containing tin metal (50-500 ppm) is streamed into air from the orifice, an oxide film is rapidly formed. The standard free energies of lead oxide and tin oxide formation are -75 and -105 kcal/mol, respectively. In the outermost layer, a sufficient amount of oxygen is available and no apparent selective oxidation between the two metals could be expected. Once the oxide film has formed, oxygen atoms have to penetrate this film to reach the inner part, so only a limited

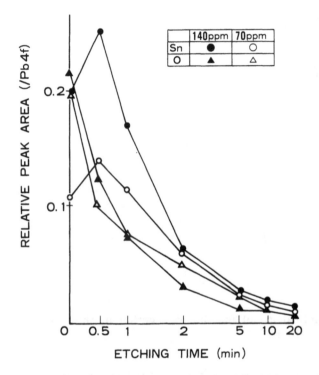

Figure 6.12 Depth profile of Sn_{3d} and O_{1s} in the tin-containing lead fibers. Relative peak area of Sn_{3d} or O_{1s} to Pb_{4f} are indicated.

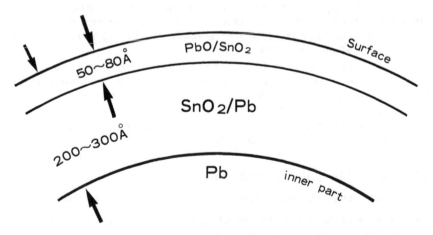

Figure 6.13 Model of the estimated surface structure of Pb/Sn fibers from XPS spectra.

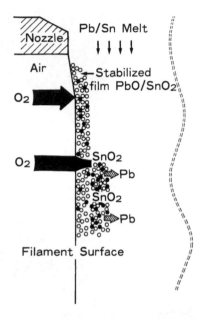

Figure 6.14 Hypothetical model for Sn segregation in the outer surface of the Pb/Sn fibers.

amount of oxygen is available for oxidation of the metals. Thus, selective oxidation of tin, rather than that of lead, can occur because of its higher oxygen affinity. The volume expansion accompanying the coagulation of tin oxide pushes molten lead to the inner part. Accordingly, the concentration of tin increases more in the outer layer than in the inner layer. In fact, the integrated intensity ratio of Sn_{3d} to Pb_{4f} is higher in a layer of several tens of angstroms inside the surface (Fig. 12).

In tin-containing lead fibers, the surface film is composed of SnO_2 and PbO parts. As the melting point of SnO_2 is higher than PbO, the modulus of the SnO_2 part is thought to be higher. Thus, the addition of tin is effective for the stabilization of the surface film and the varicose oscillation is fully prevented, affording a smooth surface on the resulting fiber.

3.4 Crystal Structure of Lead Fibers

3.4.1 Wide-Angle X-Ray Diffraction Patterns of Lead Fibers

Figures 15 and 16 illustrate the observed X-ray diffraction patterns of pure lead fibers and tin-containing lead fibers, respectively.

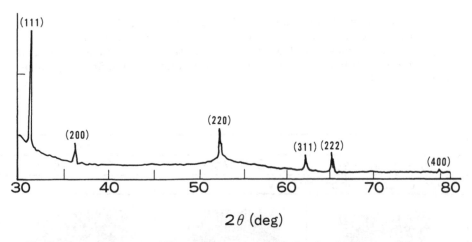

Figure 6.15 X-ray diffraction pattern of the lead fibers. The pattern was taken with Ni-filtered CuK$_\alpha$ radiation with lead fibers being arranged in a definite direction and fixed on the polypropylene thin film by colloidal solution.

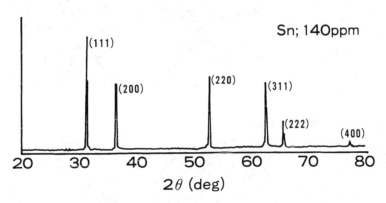

Figure 6.16 X-ray diffraction pattern of tin-containing lead fibers. The pattern was taken under the conditions described in Fig. 15.

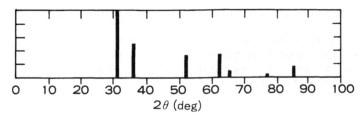

Figure 6.17 JCPDS standard diffraction pattern of Pb(f.c.c.).

The peak intensities and positions of these two X-ray diffraction patterns agree well with the JCPDS standard diffraction pattern of Pb (f.c.c.) shown in Fig. 17. Even in the tin-added lead fibers, other peaks originating from the tin compounds or Pb/Sn alloy were not observed.

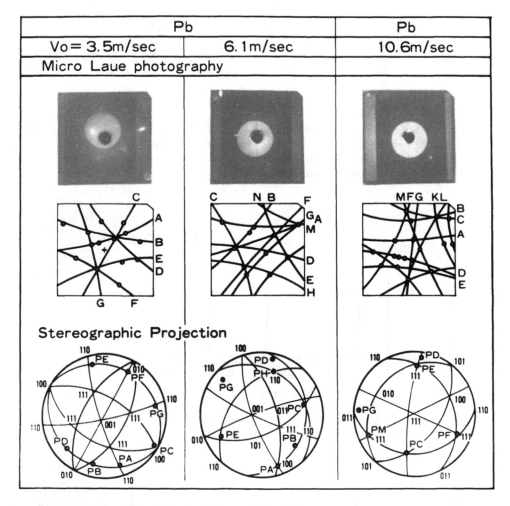

Figure 6.18 Crystal orientation of lead fibers streamed at various initial jet velocities. The Laue photographs were taken using a single filament. X-ray: CuK$_\alpha$ radiation; camera radius: 20 mm; spinning orifice diameter: 0.05 mm.

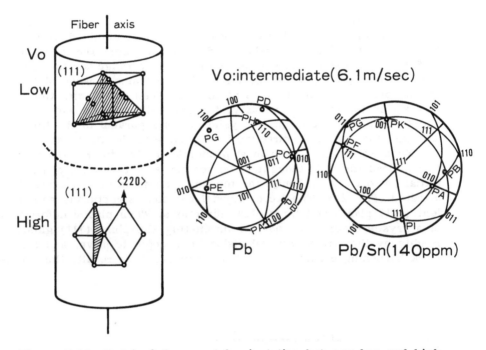

Figure 6.19 Model of the crystal orientation between low and high jet velocity. The difference in the crystal orientation between Pb and Pb/Sn fibers is compared at the intermediate velocity range of 6.1 m/sec.

Then, the crystal structures of both the pure lead fibers and tin-added lead fibers were determined as the face-centered cubic (f.c.c.) configuration.

3.4.2 Preferred Orientation of the Crystal

The orientation of the crystallites in a lead fiber is determined by the back-reflection micro-Laue method. Figure 18 shows Laue photographs of each of a single filament of pure lead fibers obtained with various initial jet velocities. Selected diffraction spots traced from the photographs and stereographic projections of crystals are also illustrated in Fig. 18. In all cases, Laue spots are clearly observed, and it is indicated that fairly large crystals are grown in the fiber texture. When the low initial jet velocity, 3.5 m/sec, is employed, (100) planes orient parallel to the fiber axis and <110> directions are perpendicular to the fiber axis. In contrast, (111) planes and <110> directions parallel to the fiber axis result when

the high initial jet velocity, 10.6 m/sec, is employed. A perturbed
orientation of the crystal is observed in the fiber obtained with
the intermediate initial jet velocity, 6.1 m/sec. Thus, the jet velocity
of 6.1 m/sec is considered to be in the velocity range for the transition of the crystal orientation. Figure 19 illustrates a model of the
crystal orientation. In the case of the pure lead fiber, with the
increase in the initial jet velocity, the crystallites rotate and the
slip (111) plane of the f.c.c. crystal becomes parallel to the fiber
axis and the slip < 110 > direction is aligned with the fiber axis.

   When we consider mechanical properties of the fibers, the elongation of lead fibers is thought to be considerably affected by the
initial jet velocity, although the tensile strength is not significantly
affected (Fig. 20). The f.c.c. crystal has four structurally equivalent (111) planes and three < 110 > directions. When plastic deformation
of a crystal occurs, the most predominating slip plane (111) glides
readily by the stress. Several dislocations begin to move, and then
work hardening occurs in the fiber textures. As a result, the elonga-

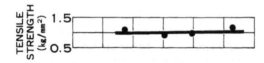

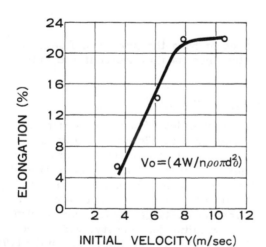

Figure 6.20   Mechanical properties of pure lead fibers. Sample length:
20 mm; draw ratio: 10 mm/min.

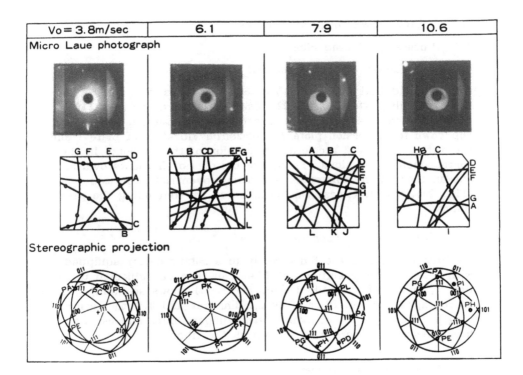

Figure 6.21 Crystal orientation of Pb/Sn (140 ppm) fibers streamed at various initial jet velocities from an orifice of 0.05 mm diameter.

tion of the pure lead fibers is estimated to become larger. On the other hand, in the case of tin-containing lead fibers, the (111) plane is aligned parallel to the fiber axis even when the intermediate velocity, 6.1 m/sec, is employed (Fig. 19). Thus, the crystal orientation as well as the elongation of the fiber is not significantly affected by the initial jet velocity. It is shown more precisely in Fig. 21 that the crystal orientation parallel to the fiber axis is more readily attained in tin-containing lead fibers compared to pure lead fibers. In the case of tin-containing lead fibers, tin oxide film is formed on the surface of the jet. The film of the surface has a high modulus and effectively supports and stabilizes the jet. Thus, total downward force on the filament originating from aerodynamic and gravitational forces is applied to the coagulating jet effectively. Consequently, the preferred orientation of the (111) plane parallel to the fiber axis is assumed to be easily attained.

4. APPLICATIONS OF LEAD FIBERS

4.1 Processing of Lead Fiber

4.1.1 *Production of Lead Fiber Mat*

As-spun lead fibers can be subsequently processed to mat by the apparatus illustrated in Fig. 22. The melt spun lead fibers (100 to 400 filaments) are extruded directly onto dispersion plate 1 and slip down to dispersion plate 2. Horizontal vibration of plate 1 and the motion of plate 2 accelerate dispersion of the filaments in the corresponding directions. Thus, the bundles of filaments overlap one another on the conveyer, and a nonwoven mat is formed (Fig. 23). The thickness can be adjusted by controlling the press roller.

4.1.2 *Production of Lead Cut Fibers*

As-spun lead fibers are fed directly to a rotary cutter equipped with a sieve. The resulting cut fiber can then be subjected to selection by sieves of a specified size. Figure 24 shows scanning electron micrographs of a cut fiber of 40 μm diameter and 0.7 mm mean length. These cut fibers exhibit good dispersibility and can be homogeneously dispersed in a resin as the filler.

4.2 Sound Control Materials

It is well known that heavy, soft materials effectively reflect the sound wave impinging on them. Lead sheet has such characteristics and has been used as a sound insulator. However, in practice,

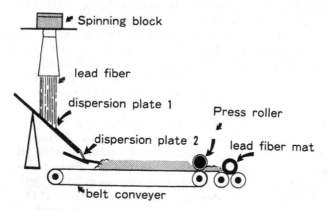

Figure 6.22   Process diagram to produce lead fibers mat.

Figure 6.23 Photograph of a lead fiber mat.

lead sheet has been laminated with supporting materials such as
plywood or slate, since it is easily torn and tends to creep when
it is hung. This characteristic sacrifices its original flexibility and
fabricability and sets a significant limitation to its actual application.

To overcome this problem, a lead fiber nonwoven fabric is em-
bedded in soft polyvinyl chloride [46]. Some examples currently
on the market (Toray's FC sheets) are shown in Table 4. Figure 25
shows the sound insulating property of #1000 in terms of transmission
loss, which is defined as

$$\text{Transmission loss (in dB)} - 10 \, \log\left(\frac{I_0}{I}\right)$$

where I_0 is the sound energy impinging on the insulation material
and I is the sound energy transmitted through the material.

The thin, flexible, and highly dense #1000 sheet does not demonstrate any reduction in its sound insulation effectiveness caused
by coincidence effect (Fig. 26), which is generally unavoidable
for hard materials such as iron plates, concrete, and plywood.

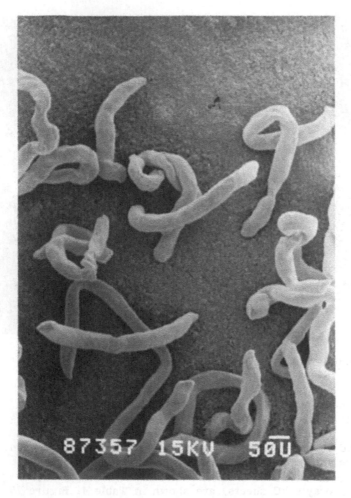

Figure 6.24 Scanning electron micrograph of lead cut fibers. Mean length about 0.7 mm.

Over the entire audible frequency range, the sound insulation capacity of #1000 surpasses predictions based on the mass law: Transmission loss = 18 log M·f - 44, where M is the surface mass (kg/m^2) of the material and f is the frequency (Hz) of the incident sound. The lead fiber-embedded sheets have also shown good vibration damping (Table 5). When the sheets are mounted over a vibrating noise source, (a) sound produced as a result of impact is reduced and (b) noise as a result of resonance or vibration is diminished.

Table 4   Sound Insulation Materials

	Weight (kg/m^2)	Thickness (mm)	Structure
#1000	3.4	1	PVC lead fiber
			PVC lead fiber
#2000	3.4	1	nonwoven fabric
			Lead fiber PVC
#7000	3.6	1	woven fabric

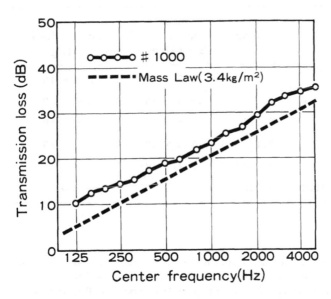

Figure 6.25 Sound insulation property of #1000 sheet.

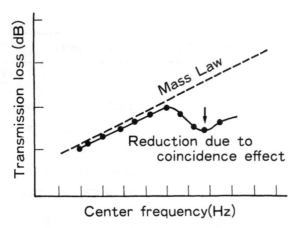

Figure 6.26   Coincidence effect.

Table 5   Damping Properties

Material	Loss factor η	Young's modulus E (kg/cm^2)	Loss modulus $\eta \times E$ (kg/cm^2)
FC #1000	0.3	9.8×10^3	2.9×10^3
FC #2000	0.3	10.0×10^3	3.0×10^3
Lead sheet	0.015	1.6×10^5	2.4×10^3
Rubber	0.1	1.0×10^2	1.0×10
Steel plate	0.00006	2.1×10^6	—
Steel plate + FC #1000	0.082	—	—

Measured at 20°C.

The effectiveness of damping materials is measured in loss modulus $\eta \times E$, where η is the loss factor and E is Young's modulus.

An application of #7000 for an engine test room is illustrated in Figs. 27 and 28. Sound reduction of as much as 27 dBA is noted.

4.3   Radiation Shielding Materials

4.3.1   γ-Ray Shielding Material

Nonwoven lead fiber mat can be used as a radiation shielding material for regular inspection and repairing at nuclear power stations.

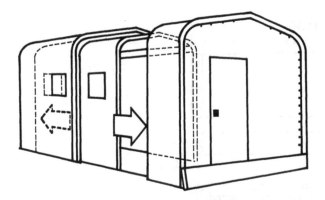

Figure 6.27 Engine test room.

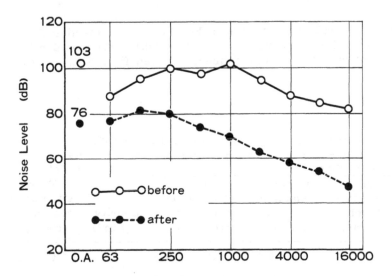

Figure 6.28 Noise reduction capacity of an engine test room.

Toray's FC-MAT, shown in Fig. 29, is a typical example of such application of the mat. The elementary shielding ability of FC-MAT is shown in Tables 6 and 7. Shielding of radiation of both γ-rays and X-rays is used in medical fields, nuclear power stations, and any other inspection facilities where radiation is involved. Besides mat, various types of product such as belt type, cylindrical type, cutting type, cutting products, head gears, and aprons are also available (Figs. 30 and 31).

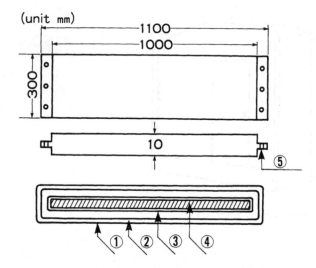

Figure 6.29 Specification of Toray's FC-MAT: 1 protective film, polyethylene (100 μm thickness); 2 outer cloth, film coated polyester; 3 inner cloth, synthetic fiber; 4 lead fiber mat; 5 eyelet.

Table 6.6 Elementary Shielding Ability for Lead Fiber Web

	Lead fiber			Lead plate	
Diameter of fiber (μm)	Bulk density of web (g/cm^3)	Linear attenuation coefficient (cm^{-1})	Mass attenuation coefficient (cm^2/g)	Density (g/cm^3)	Mass attenuation coefficient (cm^2/g)
30	3.2	^{60}Co 0.18 ^{137}Cs 0.34	^{60}Co 0.056 ^{137}Cs 0.106	11.3	^{60}Co 0.059 ^{137}Cs 0.102

Analytical data at Tokyo Metropolitan Isotope R.C.

Table 6.7 Efficiency of FC-MAT Gamma Radiation Shielding

Sheet number of FC-MAT	Attenuation of γ-ray radiation		
	^{60}Co	^{137}Cs	^{226}Ra
0	1.0	1.0	1.0
1	0.82	0.68	0.73
2	0.63	0.47	0.57
3	0.53	0.33	0.44
4	0.45	0.20	0.31

At Power Reactor & Nuclear Fuel Development Corp.

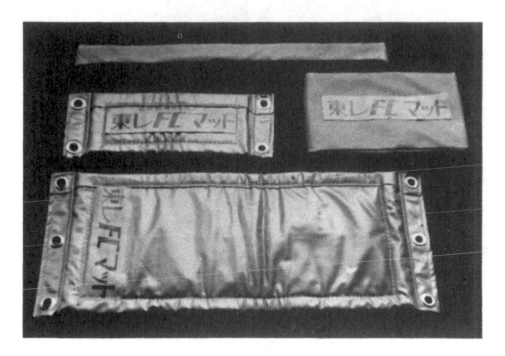

Figure 6.30 Various types of FC-MAT.

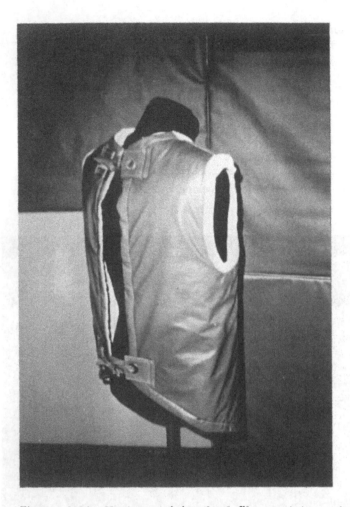

Figure 6.31 Vest containing lead fiber mat to protect workers at a nuclear power station.

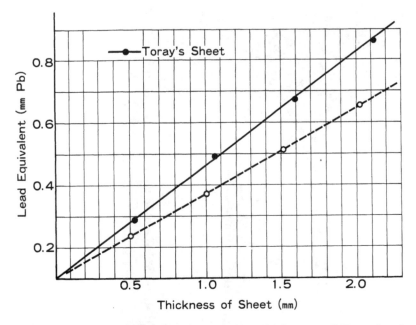

Figure 6.32 Relationship between the thickness of Toray's elementary sheet and lead equivalent. Dashed line: efficiency of ordinal X-ray shielding sheet containing lead powder (lead sulfate, lead oxide).

4.3.2 X-Ray Shielding Sheet

A blended product of fine lead cut fiber with resin has a number of advantageous properties such as the very thin thickness, high specific gravity performance, and high efficiency for shielding radiation. Conventional resin sheets containing a powder of a lead compound(s) such as lead sulfate and lead oxide are less effective because of the lower percentage of lead content. The efficiency of Toray's X-ray shielding sheet is shown in Fig. 32. Lead equivalents are 0.17 mmPb for 0.5 mm in thickness and 0.33 mmPb for 1.0 mm in thickness. The sheet exhibits excellent flexibility and is easily subjected to various processes of cutting, sewing, and trimming. Thus, the sheet would be suitable for making work cloth. Applications of the sheet include (a) X-ray shielding material for aprons, groves, partitions, curtains, etc., used in medicine (Fig. 33); (b) shielding material for nondestructive inspection; and (3) shielding material for work clothes at nuclear power stations.

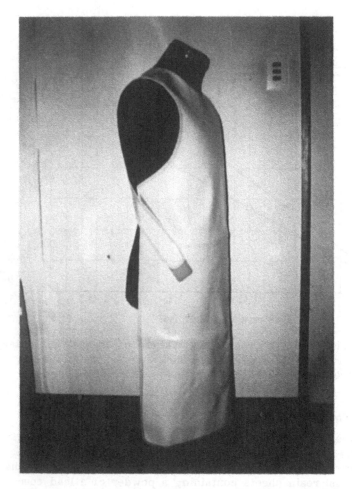

Figure 6.33 Apron made of Toray's X-ray shielding sheet for medical use.

Table 6.8 Properties of a Nylon 6/Lead Filament

Average denier (d)	1,250
Tensile strength (kg)	1.8
Elongation (%)	22
Average density (g/cm^3)	2.15

4.4 Fishing Net

Conjugated filaments consisting of a nylon 6 sheath and a lead alloy core (fractured, 2-3 mm long) have been used for fishing nets, mainly fixed nets [47]. Since the tensile strength of the filaments with lead cores is rather low (Table 8), they are intertwisted with nylon or polyester filaments into a twine. Meeting our expectation, actually increased catches of fish have been reported thus far, which may be explained by its high maintainability of the shape in the seawater resulting from its high twine density.

REFERENCES

1. U.S. Patents 2,907,082; 2,976,590.
2. N. E. Alber, Willowick and W. E. Smith, U.S. Patent 3,216,076.
3. M. Nagano, T. Sawaki, and S. Taniguchi, *Sen-i Gakkaishi*, 24 (4), 258 (1968).
4. British Patent Specifications 1,153,577 (1969), 1,197,972 (1970).
5. M. Nagano, T. Sawaki, S. Taniguchi, *Sen-i Gakkaishi*, 26 (6), 283 (1970).
6. U.S. Patents 3,583,027; 3,584,678.
7. M. Nagano, *Sen-i Gakkaishi*, 28 (1), 23 (1972).
8. I. G. Butler, W. Kurtz, J. Gillot, B. Lux, *Fiber Sci. Technol.* 5 (4), 243 (1972).
9. R. E. Cunningham, W. J. Privott, Jr., and L. F. Rakestraw, U.S. Patent 3,720,741.
10. J. C. Hubert, F. Mollard, and B. Lux, *Z. Metallkde.*, 64 (12), 835 (1973).
11. D. J. Thorne, *Fiber Sci. Technol.*, 7, 79 (1974).
12. T. Sawaki, *J. Text. Mach. Soc. Jap.*, 21 (2), 35 (1975).
13. K. Tanno, F. Takahashi, and S. Kikuta, *Sen-i Gakkaishi*, 33 (2), T-63 (1977).
14. I. Ohnaka and T. Fukusako, *J. Jap. Inst. Metals*, 42 (4), 415 (1978).
15. S. Kavesh, *AIChE Symp. Ser.*, 74 (180), 1 (1978).
16. R. F. Cunningham, L. F. Rakestraw, and S. A. Dunn, *AIChE Symp. Ser.* 74 (180), 20 (1978).
17. J. M. Massoubre and B. F. Phlieger, *AIChE Symp. Ser.* 74 (180), 48 (1978).
18. R. E. Maringer and C. E. Mobley, *AIChE Symp. Ser.*, 74 (180), 16 (1978).
19. E. W. Collings, C. E. Mobley, and R. E. Maringer, *AIChE Symp. Ser.*, 74 (180), 102 (1978).

20. R. E. Maringer and C. E. Mobley, Rapidly Quenched Metals in Proc. Int. Conf., 3RD P49-56, 1978.
21. J. V. Wood and S. C. King, *J. Mater. Sci.*, *13*, 1119 (1978).
22. R. B. Pond, Sr., and J. M. Winter, Jr., *AIChE Symp. Ser.* *74* (180), 98 (1978).
23. G. F. Taylor, *Phys. Rev.*, *23*, 655 (1924).
24. U.S. Patent 1,793,529.
25. H. Wagner, *Wire J.*, *6*, 871 (1961).
26. J. Nixdorf and H. Rochow, *Battelle-Information*, *3*, 4 (1968).
27. J. Nixdorf, *Proc. Roy. Soc. London A*, *319*, 17 (1970).
28. *Composite*, *1*, 167 (1970).
29. M. Nagano, Y. Yuki, T. Goto, and R. Ide, *Sen-i Gakkaishi*, *29*, T-461 (1973).
30. G. Manfre, G. Servi, and C. Ruffino, *J. Mater.Sci.*, *9*, 74 (1974).
31. T. Goto et al., *Trans. Jap. Inst. Met.*, *18*, 207, 557, 562, 759 (1970).
32. T. Goto, *Sen-i Gakkaishi*, *34* (6), T-237 (1978).
33. G. W. F. Pardoe, E. Buttler, and D. Gelder, *J. Mater. Sci.*, *13*, 786 (1978).
34. T. Goto, *Trans. Jap. Inst. Met.*, *19*, 60, 605 (1978).
35. T. Goto, *Trans. Jap. Inst. Met.*, *22* (2), 96 (1981).
36. T. Goto, *Trans. Jap. Inst. Met.*, *22* (11), 753 (1981).
37. R. P. Grant and S. Middleman, *AIChE J.*, *12* (4), 669 (1966).
38. U.S. Patent 4,485,838.
39. Japanese Patent 982,657.
40. U.S. Patent 3,001,265.
41. U.S. Patent 3,003,223.
42. Japanese Patent 903,281.
43. Japanese Patent 907,968.
44. Japanese Patent 1,073,803.
45. K. S. Kim, T. J. O'Leary, and N. Winograd, *Anal. Chem.*, *45* (13), 2214 (1973).
46. Japanese Patent 975,848.
47. Japanese Utility Model 1,380,940.

7

POLYSTYRENE-BASED FUNCTIONAL FIBERS

MASAHARU SHIMAMURA, KAZUO TERAMOTO,* TOSHIO
YOSHIOKA, and MICHIHIKO TANAKA / Toray Industries, Inc.,
Otsu-shi, Shiga, Japan

*Present affiliation: Biomaterial Research Institute Co. Ltd.,
Yokohama, Kanagawa, Japan

1. INTRODUCTION

Recently, a variety of new fiber materials have been developed
and have found industrial applications based on their excellent
mechanical properties and/or special functions. Such fiber materials
are often called "specialty fibers" and have become the new family
of synthetic fibers. One of the most important groups of specialty
fibers is fibers having a remarkable durability against various
environmental changes. Aramid fiber and graphite fiber are typical
examples of this group, characterized by their high tensile strength
and high modulus, together with their heat resistivity. Another
important group is fibers having a new and special function(s),
such as optical fibers for signal transmission, hollow fibers for
gas or liquid separation, organic conductive fibers, and active
carbon fibers. Polystyrene-based functional fibers discussed in
this chapter also fall in the latter group.

Polystyrene has been widely used as the basic polymer for
a number of functional polymers, such as ion exchangers, most
of which are available in a bead form but not in a fibrous form.
However, fiber is generally considered to be one of the most practical and suitable forms for functional polymers to display their own
special functions. Especially because of the large specific surface
area, a fibrous form is expected to offer a great advantage when
the solid surface of the polymer plays an important role in the
mechanism of exhibiting a special function, as is the case of ion
exchangers. Furthermore, a wide variety of well-developed fiber
technologies, including spinning, weaving, knitting, as well as
paper making, can be effectively applied to draw upon and/or
enhance the special functions of the polymers. However, it is well
recognized that a fibrous polystyrene exhibits only poor mechanical
strength.

In fact, a fibrous polystyrene is brittle and readily breaks
into small pieces. Also, a fibrous polystyrene is often not durable
to reactions established for introduction of functional groups to
a solid state, bead-type polystyrene. In order to obtain a fibrous
polystyrene with enhanced mechanical strength, a blend with poly-

propylene has occasionally been employed, at the expense of sacrificing a considerable area covered with the undesired component, polypropylene. In our laboratory, polystyrene has been investigated in an attempt to obtain a polystyrene fiber with a reasonably high mechanical strength. As a result, a polystyrene fiber with an islands-in-a-sea structure has recently been developed. This polystyrene fiber exhibits an acceptably high mechanical strength and can be subjected to a wide variety of chemical reactions. Therefore, using this fiber as the common precursor, a variety of polystyrene-based functional fibers can be prepared by introducing an appropriate functional group. The purpose of this chapter is to review the preparation methods, physical and chemical properties, and characteristics of polystyrene-based functional fibers, together with their applications.

2. PREPARATION

Approaches to preparation of functional fibers are essentially of three different types: (a) synthesis of a functional polymer followed by spinning into a fiber; (b) functionalization by special spinning techniques; and (c) fiber preparation followed by introduction of functional groups.

Our approach to preparation of polystyrene-based functional fibers can be classified at type (c). Since a number of chemical reactions have been established for introduction of functional groups, a type (c) approach offers a great advantage in that a variety of functional fibers can readily be prepared from one common precursor fiber by employing such reactions. However, approaches of type (c) require polystyrene fibers with an acceptably high mechanical strength as a prerequisite. Our approach to improving polystyrene fiber is adoption of an additional fiber-forming polymer for reinforcement. To obtain a better reinforcing effect, the additional polymer is incorporated as the island component of an islands-in-a-sea type of composite fiber. The resulting composite fiber can be further subjected to crosslinking reaction of the sea component, polystyrene, to an adequate extent. From these composite fibers, a variety of polystyrene-based functional fibers can be prepared.

2.1 Polystyrene-Polypropylene Composite Fiber

For preparation of polystyrene-polypropylene composite fiber, 40 parts of polystyrene and 10 parts of polypropylene (as the sea ingredient) and 50 parts of polypropylene (as the island ingredient) are melt spun at 255°C into composite filaments with an islands-in-a-sea type of sectional structure as shown in Fig. 1. The composite

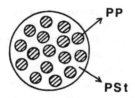

PP

PSt

Figure 7.1 Schematic cross section of composite fiber. Pst: poly-styrene, sea ingredient; PP: polypropylene, island ingredient.

filaments can be drawn to an adequate extent in the conventional manner. The diameters of the sea and the islands of the resulting filaments are ca. 24 and 4.2 μm, respectively. Their tensile strength is around 3.0 g/d.

Polystyrene is used as the basic polymer for the introduction of functional groups, while polypropylene, which exhibits an en-hanced resistance to chemicals, is used as the fiber-forming polymer for reinforcement. Polypropylene is also used as one of the sea ingredients to increase the affinity between the sea and island parts and to prevent the separation of the two phases. The composite fiber in various forms, such as filament, cut fiber, knitted fabric, woven fabric, braid, felt, nonwoven fabric, chip, and paper, can be obtained by processing the filaments, as the filaments reinforced with polypropylene have a reasonably high mechanical strength. Figure 2 shows the polystyrene-polypropylene composite fiber in the various forms.

Polystyrene-based ion exchange fibers, named IONEX, are pre-pared according to the process shown in Fig. 3 after spinning [1].

2.2 Crosslinked Polystyrene Fiber

Polystyrene-polypropylene composite fiber can be crosslinked in the presence of para-formaldehyde. In a typical reaction, 10 g of the fiber are crosslinked in 100 ml of a solution consisting of 5 wt% para-formaldehyde, 25 (45) wt% acetic acid, and 70 (50) wt% concentrated sulfuric acid. The degree of crosslinkage depends on reaction temperature as well as on reaction period. The methods of chemical reaction to the polystyrene part are shown in Fig. 4.

2.3 Chloromethylated Polystyrene Fiber

Chloromethylation of the crosslinked fiber is carried out using chloro-methyl methyl ether and the catalyst, tin chloride. Typical conditions are 10 g of the crosslinked fiber, 100 ml of chloromethyl methyl ether,

(a) **(b)** **(c)**

(d) **(e)** **(f)**

Figure 7.2 Polystyrene-polypropylene composite fiber in various forms of (a) filament, (b) needle punched felt, (c) knitted fabric, (d) cut fiber, (e) chip, and (f) paper.

Figure 7.3 Process of preparing the polystyrene-based ion exchange fiber (IONEX).

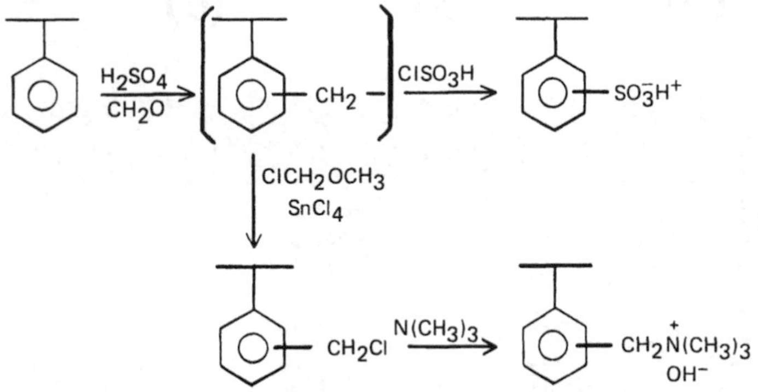

Figure 7.4 Methods of chemical reaction to polystyrene part.

10 ml of tin(IV) chloride, temperature of 30°C and reaction period of 1 hr. The chloromethylated fiber is very reactive and therefore can be used as a basic fiber for the introduction of various functional groups described below.

2.4 Cation Exchange Fiber

A strong cation IONEX is obtained by the sulfonation of the cross-linked fiber (10 g) in a solution comprising 200 ml of trichloroethylene and 8 ml of chlorosulfonic acid at 20°C for 2 hr. Figures 5 and 6 present a strong cation IONEX in the form of cut fibers and other forms, respectively.

2.5 Anion Exchange Fiber

Various anion IONEX with aminomethyl, secondary and tertiary amino, and quaternary ammonium groups can readily be prepared by treating the chloromethylated fiber with corresponding amine solution.

A strong anion IONEX is obtained by the quaternization of the chloromethylated fiber (10 g) in 100 ml of a 30 wt% trimethylamine aqueous solution at 30°C for 1 hr.

2.6 Chelating Fiber

The iminodiacetic acid (IDA) type of chelating IONEX is obtained by treating the aminomethylated fiber (10 g) with a solution consisting of 27 g of chloroacetic acid, 11 g of sodium hydroxide, 19 g

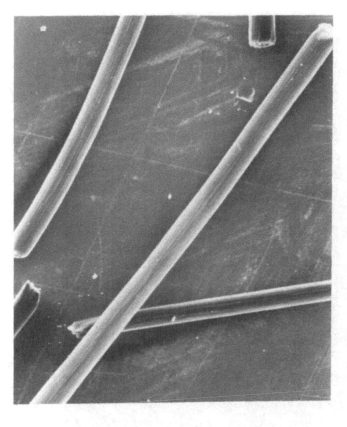

100 μm

Figure 7.5 Strong cation IONEX in a form of cut fibers.

of sodium carbonate, and 130 g of water at 100°C for 8 hr [2].
On the other hand, 10 g of chloromethylated fiber are treated in
a solution consisting of 160 g of N,N,N",N"-tetrakis(2-cyanoethyl)
diethylenetriamine and 350 ml of dioxane at 70°C for 6 hr. Then
the iminodipropionic acid (IDP) type of chelating IONEX is obtained
by treating the fiber with 450 ml of concentrated hydrochloric acid
at 60°C for 6 hr [2]. The structures of the IDA and IDP chelating
groups are shown in Fig. 7.

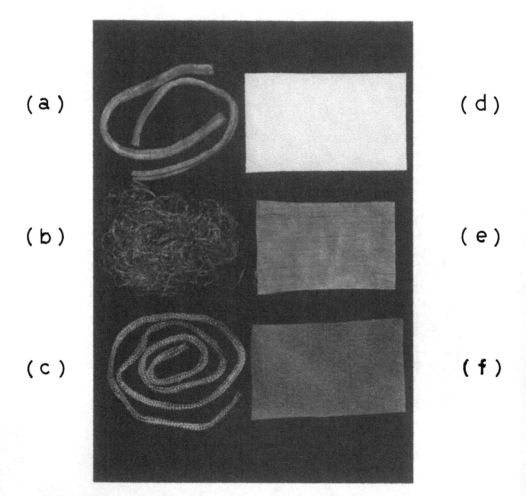

Figure 7.6 Strong cation IONEX in various forms of (a) knitted cord, (b) filament, (c) braid, (d) paper, (e) knitted fabric, and (f) needle punched felt.

2.7 Fiber-Supported Quaternary Phosphonium Ion and Crown Ether

The quaternary phosphonium ion and crown ether supported on the polystyrene-polypropylene composite fiber (see Fig. 8) have easily been prepared by treating the chloromethylated fiber in organic solvents containing tributhylphosphine and monoaza-18-crown-6, respectively [3].

(a)

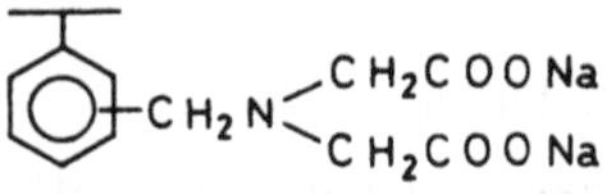

$\quad$ —CH$_2$N$\overset{\displaystyle \diagup CH_2COONa}{\diagdown CH_2COONa}$

(b)

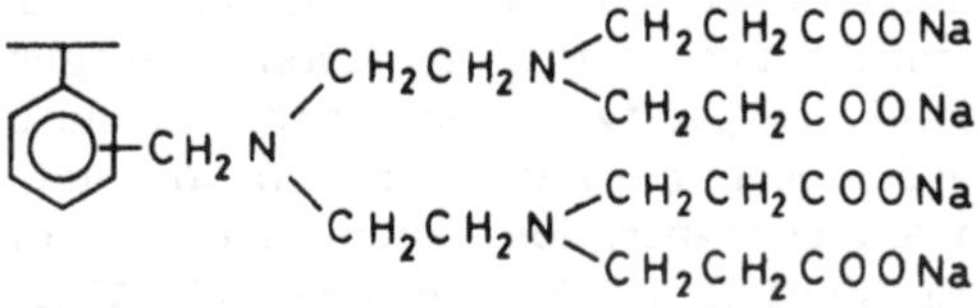

$\quad$ —CH$_2$N$\overset{\displaystyle \diagup CH_2CH_2N\overset{\displaystyle \diagup CH_2CH_2COONa}{\diagdown CH_2CH_2COONa}}{\diagdown CH_2CH_2N\overset{\displaystyle \diagup CH_2CH_2COONa}{\diagdown CH_2CH_2COONa}}$

Figure 7.7 Structures of the iminodicarboxylic acid chelating groups: (a) IDA-Na, (b) IDP-Na.

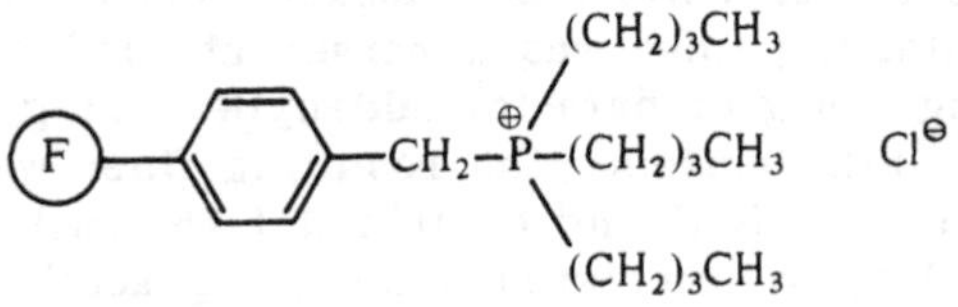

F —$\bigcirc$— CH$_2$—$\overset{\oplus}{P}$—(CH$_2$)$_3$CH$_3$ $\quad$ Cl$^{\ominus}$

with (CH$_2$)$_3$CH$_3$ above and (CH$_2$)$_3$CH$_3$ below

(a)

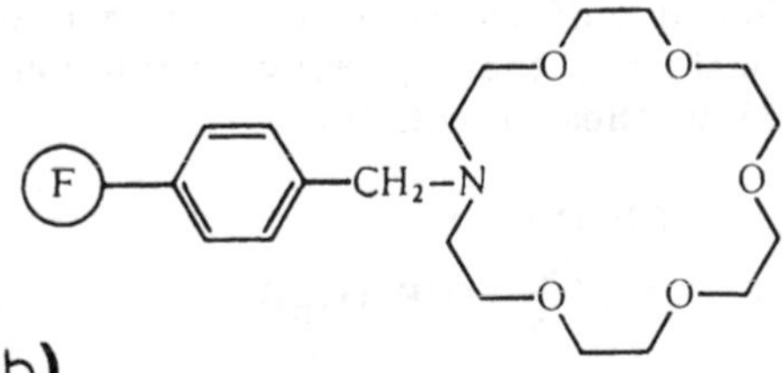

F —$\bigcirc$— CH$_2$—N (crown ether ring)

(b)

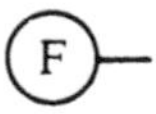

F — : Fibrous polymeric support

Figure 7.8 Structures of (a) fiber-supported quaternary phosphonium ion and (b) crown ether.

2.8   New Chemical Reactions

2.8.1   Crosslinking and Hydroxymethylation

Polystyrene fiber can also be crosslinked by treatment with a solution containing para-formaldehyde, sulfuric acid, and nitrobenzene at room temperature [4]. The resulting crosslinked polystyrene fiber is insoluble. In this reaction, the para-formaldehyde, the sulfuric acid, and the nitrobenzene serve as the crosslinking reagent, catalyst, and swelling reagent, respectively. Any polystyrene or its reaction products released from the polymer matrix are not observed in the solution even in low concentration, such as 0.03%, of para-formaldehyde (para-formaldehyde 0.035%, sulfuric acid 18%, and nitrobenzene 82%). These results suggests that the crosslinking reaction proceeds much faster than the dissolution of the polystyrene.

When the concentration of para-formaldehyde is high, the weight of the fiber increases 15-30% of that of the starting fiber of which polystyrene content is 50%. This weight-increased fiber contains 1 mole of hydroxyl group per 3 to 4 moles of aromatic nuclei. For example, when 1 g of the starting fiber was immersed at 20°C for 5 hr in a solution containing 5.6 g of para-formaldehyde, 7.1 g of sulfuric acid, and 33.4 g of nitrobenzene, the resulting fiber was determined to be 1.166 g in its weight and contained 1.45 mmol of hydroxyl group as estimated by volumetric analysis using acetic anhydride in pyridine. And the greater part (93.9%) of the resulting fiber was insoluble in boiling toluene (1 g of the fiber per 100 ml of toluene, boiling for 2 hr).

The higher the concentration of sulfuric acid, the fewer hydroxyl groups are introduced. It is considered that hydroxymethylation (1) occurs with crosslinking (2) in these reactions.

$$-CH_2CH- \qquad + (CH_2O)_m \rightarrow \qquad -CH_2CH- \quad (CH_2O)_nH \qquad (1)$$

$$2 \quad -CH_2CH- \quad + (CH_2O)_m \rightarrow \quad -CH_2CH- \qquad -CH_2CH- \quad CH_2 \qquad (2)$$

The larger the mixing ratio of nitrobenzene to sulfuric acid, the higher the rate of crosslinking reaction, though the concentration of the catalyst decreases. These results suggest that the swelling is the rate-determining step in the crosslinking reaction.

Besides sulfuric acid, methane sulfonic acid and laurylbenzene sulfonic acid are available as catalysts, but their activities are much lower than that of sulfuric acid. As swelling agent, nitropropane is also usable, but in this case, para-formaldehyde should be used at about ten times higher concentration than in the case of nitrobenzene in order to prevent the release of the polystyrene from the fiber matrix. As for the other reaction conditions, room temperature is adopted to avoid sulfonation.

2.8.2 Amidomethylation

The crosslinked polystyrene fiber is acetamidomethylated by treating the mixed solution with N-methylol acetamide, sulfuric acid, and nitrobenzene (1:6.5:8.4 by weight) at 15-20°C, and the acetamido-methylated fiber is converted to the aminomethylated one by hydrolysis (3).

$$-CH_2CH- \qquad \qquad -CH_2CH-$$

$$+ CH_3CONHCH_2OH \longrightarrow \bigcirc\!\!\!\!\!+\!\!-CH_2NHCOCH_3$$

$$\xrightarrow[HCl]{H_2O} \quad -CH_2CH- \qquad \qquad (3)$$

$$\bigcirc\!\!\!\!\!+\!\!-CH_2NH_2 \cdot HCl$$

As an amidomethylating solution is labile to both temperature and moisture, it should be prepared at a temperature below 20°C. As the reaction rate is very fast, the reaction goes practically to completion within 5 hr. However, when the polystyrene fiber is highly crosslinked, it hardly undergoes the amidomethylation due to the lack of both swelling ability of the fiber and stability of the reacting solution, even if the reaction time is prolonged.

Besides, N-methylol acetamide and N-methylol compounds of stearamide and benzamide are available, but the resulting fiber is so hydrophobic that hydrolysis of the amide group does not proceed. Therefore, these N-methylol compounds are not suitable for preparing an aminomethylated polystyrene fiber. In this reaction, ester and ether derivatives of N-methylol amides are also available as the amidomethylating reagent. Sulfuric acid is the best catalyst and also acts as solvent.

Nitropropane as well as nitrobenzene is a good solvent, when sulfuric acid is used as catalyst.

2.8.3 One-Step Crosslinking—Amidomethylation

Both crosslinking and amidomethylation are due to electrophilic substitution reactions of electrophiles to aromatic nucleus. Therefore, they can be allowed simultaneously. In this reaction system, swelling-dissolving, crosslinking, and amidomethylation of polystyrene proceed competitively. When the concentration of para-formaldehyde is too high, the polystyrene is hardly amidomethylated. On the contrary, when it is too low, it begins to dissolve into the solution [5].

Therefore, the mixing ratio of these reagents is very important. To prevent the release of polystyrene from the fiber matrix, the concentration of para-formaldehyde must be higher in this reaction than in the crosslinking reaction described above. This is considered to be because, once the aromatic nucleus reacts with para-formaldehyde or N-methylol compound, it does not undergo the further substitution reaction.

To complete the reaction, N-methylol compound is needed in large excess. For example, when one part of the fiber was immersed in a solution containing para-formaldehyde, N-methylol acetamide, sulfuric acid, and nitrobenzene at a 0.054-0.9-9.8-20 or 0.10-3.1-25-26 ratio by weight at room temperature at 15 hr, the theoretical amount of amidomethyl group was introduced.

N-Methylol acrylamide is more useful as the amidomethylating reagent than N-methylol acetamide, because it is produced commercially and can easily be purified by recrystallization. Though it tends to resinify upon exposure to air at ordinary temperatures, it can be kept in a refrigerator for a long time and remains stable even in sulfuric acid below 20°C for a short time. By use of 1.5- to 2.0-fold moles excess of N-methylol acrylamide, almost all of the aromatic nuclei in the matrix are acrylamidomethylated (4).

$$-CH_2CH- \qquad\qquad + HOCH_2NHCOCH=CH_2 \quad \rightarrow \quad -CH_2CH- \qquad -CH_2NHCOCH=CH_2$$

(4)

For example, when 1 g of the polystyrene fiber was immersed in a solution comprising 0.054 g of para-formaldehyde, 1.0 g of N-methylol acrylamide, 9.6 g of sulfuric acid, and 20.0 g of nitrobenzene at room temperature for 5 hr, the completely acrylamidomethylated fiber was obtained.

The acrylamidomethyl group introduced is very reactive and is easily converted to an amino group by treating with amine. This reaction, named Michael addition, is endothermic and proceeds at room temperature. The resulting fiber is usable as a weak anion exchange fiber [6].

N-methylol-α-chloroacetamide is also useful as the amidomethylating reagent. For example, when 1 g of polystyrene fiber was dipped in the solution comprising 0.026 g of para-formaldehyde, 1.82 g of N-methylol-α-chloroacetamide, 13 g of sulfuric acid, and 13 g of nitrobenzene at 15-20°C for 2 hr, it was almost completely amidomethylated (5) [7].

$$-CH_2CH- \qquad \qquad -CH_2CH-$$
$$\bigcirc \quad + \ HOCH_2NHCOCH_2Cl \ \rightarrow \quad \bigcirc \!\!\!-\!\!\!\mid\!\!\!-CH_2NHCOCH_2Cl$$

$$(5)$$

The introduced α-chloroacetamidomethyl group reacts with ammonia and primary, secondary, or tertiary amine, being converted to the corresponding primary, secondary, or tertiary amino group or quarternary ammonium acetamidomethyl group. In the reaction with tertiary amine, the chloroacetyl group should be converted to the iodoacetyl group, because the latter reacts with tertiary amine in milder conditions than the former. This conversion can be achieved by heating the chloroacetyl group-containing fiber in an aqueous alcohol solution of potassium iodide at 55°C for about 2 hr (6-8).

$$-CH_2CH- \qquad \qquad -CH_2CH-$$
$$\bigcirc \!\!\!-\!\!\!\mid\!\!\!-CH_2NHCOCH_2Cl \ + \ HN\!\!<^{R_1}_{R_2} \ \rightarrow \quad \bigcirc \!\!\!-\!\!\!\mid\!\!\!- \ CH_2NHCOCH_2N\!\!<^{R_1}_{R_2}$$

$$(6)$$

$$-CH_2CH- \qquad \qquad -CH_2CH-$$
$$\bigcirc \!\!\!-\!\!\!\mid\!\!\!- CH_2NHCOCH_2Cl \ + \ KI \ \rightarrow \quad \bigcirc \!\!\!-\!\!\!\mid\!\!\!-CH_2NHCOCH_2I$$

$$(7)$$

$$-CH_2CH- \qquad \qquad -CH_2CH-$$
$$\bigcirc \!\!\!-\!\!\!\mid\!\!\!- CH_2NHCOCH_2I \ + \ N\!\!<^{R_1}_{\substack{R_2 \\ R_3}} \ \rightarrow \quad \bigcirc \!\!\!-\!\!\!\mid\!\!\!- CH_2NHCOCH_2\overset{\oplus}{N}\!\!<^{R_1}_{\substack{R_2 \\ R_3}} \ I^{\ominus}$$

$$(8)$$

2.8.4 Hydrolysis of Amidomethyl Group

Almost all of the amidomethylated polystyrene fiber can be converted to the aminomethylated polystyrene fiber by heating in 6 N sulfuric acid or 6 N hydrochloric acid (9). The resulting fiber can further be converted to the dimethylaminomethylated polystyrene fiber by heating in a solution of formalin and formic acid at 90°C for 2 hr (10) and can also be converted to the IDA-type chelating fiber by treating with monochloroacetic acid in the presence of sodium carbonate, as described above (11).

$$-CH_2CH- \qquad \qquad -CH_2CH-$$
$$\bigcirc \!\!\!-\!\!\!\mid\!\!\!- CH_2NHCOR \ \xrightarrow{\ H_2O\ } \quad \bigcirc \!\!\!-\!\!\!\mid\!\!\!- CH_2NH_2 \ + \ HOCOR$$

$$(9)$$

$$-CH_2CH- \qquad \xrightarrow{CH_2O/HCOOH} \qquad -CH_2CH-$$

$$\text{(ring)}-CH_2NH_2 \qquad\qquad\qquad \text{(ring)}-CH_2N\underset{CH_3}{\overset{CH_3}{\diagdown}} \qquad (10)$$

$$-CH_2CH- \qquad \xrightarrow{ClCH_2COOH} \qquad -CH_2CH-$$

$$\text{(ring)}-CH_2NH_2 \qquad\qquad\qquad \text{(ring)}-CH_2N(CH_2COOH)_2$$

$$(11)$$

3. PHYSICAL AND CHEMICAL PROPERTIES

The strength, elongation, ion exchange capacity, and water content
of several filamentary IONEX are summarized in Table 1. The ion
exchange capacities of IONEX Nos. 1-4 corresponded to 94%, 88%,
86%, and 82% of the theoretically calculated values, respectively.
When a crosslinking group and an ionic group are introduced, the
tensile strength of the resulting filaments becomes lower than the
value estimated in consideration of the weight increase from the
data on the original filaments; that is, the tensile strength per
filament of the original filament is 12.9 g, and those of the crosslinked filament, the chloromethylated filament, and the strong anion
exchange filament are 12.0, 11.9, and 10.5 g, respectively. No
change in the capacity or water content is recognized, even if the
strong cation or anion IONEX is used 10 times in the following treatment: add to 1 N hydrochloric acid or sodium hydroxide, wash with
deionized water, add to 1 M sodium chloride, and finally wash with
deionized water. From these results, it can be presumed that a
methylene bond is formed between phenyl rings by the crosslinking
reaction and that the resulting filament is strongly crosslinked
and insolubilized. IONEX with an arbitrary water content can be
obtained by choosing the conditions of the crosslinking reaction.
The water content means the water-holding capacity of the fiber
[2]. The acidity or basicity of a strong cation or anion IONEX
is almost equal to that of polystyrene-based ion exchange resin.
    The characteristic data of the chelating IONEX are summarized
in Table 2. The chelating filaments have a diameter of ca. 30 μm
and exhibit a tensile strength of ca. 1.0 g/d. It is found that the
low-crosslinked fibers have larger adsorption capacities and water
contents than the high-crosslinked ones. The former result is explained by the fact that the amount of the aminomethyl or chloro-
methyl group introduced in the synthetic process is larger in the
case of the low-crosslinked fiber because of its high reactivity.
    The quarternary ammonium methyl group I attached to the
aromatic nucleus is stable to acid and alkali. On the other hand,

Table 7.1 Fundamental Properties of IONEX

No.	Composition of original fiber[a]	Ionic group	Diameter (μm)	Strength (g/d)	Elongation (%)	Capacity[b] (mEq/g)	Water[c] content
—	—	—	24	3.0	40	—	—
1	(40/10)/50	$-SO_3^-Na^+$	30	1.3	18	2.6	1.6
2			32	1.4	29	2.4	1.9
3	(30/7.5)/62.5	$-CH_2N^+(CH_3)_3$	32	2.0	38	1.9	1.7
4	(20/5)/75	Cl^-	28	3.0	28	1.3	1.0

[a]Sea(PSt/PP)/Islands (PP).
[b]Capacity of the strong acid (base) groups.
[c]$(W - W_0)/W_0$, where W = wet weight, W_0 = dry weight.

Table 7.2 Characteristics of Chelating IONEX

No.	Chelating group	Fiber form[a]	Degree of crosslinking	Capacity[b] (mEq/g)	Water[c] content
1		Filament	Low	2.2	2.6
2	IDA-Na	Filament	High	1.8	1.4
3		Felt	High	1.8	—
4		Filament	Low	1.5	1.7
5	IDP-Na	Filament	High	1.4	1.0

[a]Diameter and composition of original fiber: 20 μm, Sea(PSt/PP)/ Islands (PP) = (49/12)/39.
[b]Adsorption capacity for Cu^{2+}.
[c]$(W - W_0)/W_0$, where W = wet weight, W_0 = dry weight.

the quarternary ammonium acetamidomethyl group II is supposed to be labile to hydrolysis, though stable to heat. In practice, however, the functional group II in the fiber is unexpectedly stable to hydrolysis, especially when R in formula II is a longer alkyl chain than an ethyl group. Actually, the fiber withstands the conditions of steam sterilization (120°C, 30 min). Primary, secondary, and tertiary aminoacetamide group III in the fiber are more stable to heat, acid, and alkali than the functional group II.

$$-CH_2CH- \quad\quad -CH_2CH- \quad\quad -CH_2CH-$$

(I) (II) (III)

4. FUNDAMENTAL CHARACTERISTICS

4.1 Exchange of Metal Ions

The exchange of metal ions has been investigated by the method as follows [1]. A strong cation exchanger of 6.6 mEq in the Na form is added to 250 ml of a 0.008 M $CaCl_2$ aqueous solution containing 50 wt% sucrose; it is then strongly shaken at 20°C. Similarly, a strong cation exchanger of 5.7 mEq in the Ca form is added to 250 ml of a 1 M NaCl aqueous solution, and then the mixture is strongly shaken. The fractional attainment of equilibrium is evaluated in those two cases by measuring the calcium ion concentration in

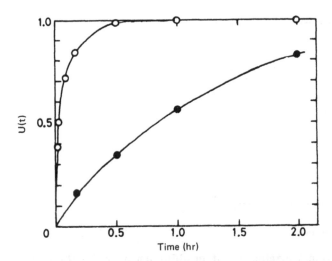

Figure 7.9 Fractional attainment of equilibrium as a function of time. Ion exchange between Na form of cation exchanger (6.6 mEq) and 0.008 M CaCl solution containing 50 wt% sucrose (250 ml). ○: IONEX No. 1 (2.5 g); ●: Amberlite IR-120B (1.5 g).

the solution at several intervals. The results are shown in Figs. 9 and 10, respectively. A commercially available strong cation exchange resin and strong cation IONEX No. 1 are used as cation exchangers.

According to the rate laws of ion exchange in a simple system, the half times, $t_{1/2}$, of ion exchange for a spherical resin with a radius of r_0 have been described in the two limiting cases of ideal film and ideal particle diffusion control, respectively, by the following equations [8]:

$$t_{1/2} = 0.230 \frac{r_0 \, \delta \overline{C}}{DC} \tag{12}$$

$$t_{1/2} = 0.030 \frac{r_0^2}{\overline{D}} \tag{13}$$

The significance of the symbols in the above equations has been described in Ref. 8.

By the modification of the method used for a spherical resin, we can obtain the half times for an endless fiber with a radius of r_f in the two limiting cases, respectively, as follows [1]:

$$t_{1/2} = 0.345 \frac{r_f \delta \overline{C}}{DC} \tag{14}$$

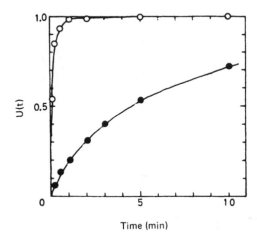

Figure 7.10 Fractional attainment of equilibrium as a function of time. Ion exchange between Ca form of cation exchanger (5.7 mEq) and 1 M NaCl solution (250 ml). ○ : IONEX No. 1 (2.2 g); ● : Amberlite IR-120B (1.3 g).

$$t_{1/2} = 0.065 \; \frac{r_f^2}{\overline{D}_f} \tag{15}$$

where $\overline{D}_f$ is the interdiffusion coefficient in the ion exchange fiber.

The results in Fig. 9 are considered to correspond to the case of film diffusion control, since the concentration of the solution is sufficiently low: C << 0.1 M. The ratio of the half time of the resin to that of the fiber can be calculated to be 11 from Eqs. (12) and (14) when r_0 = 250 μm and r_f = 15 μm. The ratio obtained from the experimental data is, however, 25, almost twice as high as the calculated ratio. This fact may be interpreted as being due to film diffusion control in terms of the increase in the surface area per unit of capacity of IONEX by the amount of the polymer for reinforcement. On the other hand, the results in Fig. 10 are considered to correspond to the case of particle diffusion control, since the concentration of the solution is sufficiently high: C >> 0.1 M; the ratio can similarly be calculated to be 128 from Eqs. (13) and (15) if $\overline{D}$ = $\overline{D}_f$ is assumed. The ratio obtained from the experimental data, however, is in the range of from 50 to 70 and is about half the calculated ratio. The results can be interpreted on the assumption that the polymer for reinforcement disturbs the diffusion of ions because of the particle diffusion control.

It can be concluded that the fiber has a high ion exchange rate for metal ions, compared with the resin.

4.2   Adsorption of Macromolecular Ionic Substances

The adsorption of a caramel pigment has been investigated by the
following procedures [1]. Two grams of a strong anion exchanger
in the Cl form is added to 200 ml of a 50 wt% sucrose aqueous solution containing a commercially available caramel pigment (C.V. 0.92);
the mixture is then stirred at 60°C. The data of the decoloration
ratio as a function of time are shown in Fig. 11. It is apparent
that the decoloration rate by the fiber is much greater than that
by the three resins. This indicates that the effective capacity of
the fiber to adsorb the caramel pigment is much larger than that
of the resins, in spite of the fact that the ion exchange capacity
of the fiber is about two-thirds that of the resins.

Next, a strong anion exchanger in the Cl form is added to
25 ml of the same sucrose solution, and the mixture is shaken at
60°C for 6 hr. The correlation between the amounts of various
anion exchangers and the decoloration ratio is shown in Fig. 12.
This shows that the fibers have from 7 to 10 times as large an
effective capacity for adsorbing the pigment as the low-crosslinked
and macroreticular (MR) types of resins and, further, 40 times
as large a capacity as the standard-crosslinked type of resin. No
large difference in the effective capacity is observed between the
strong anion IONEX Nos. 2-4. These results can be explained

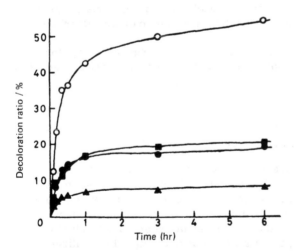

Figure 7.11   Decoloration ratio as a function of time. Decoloration
of 50 wt% sucrose solution containing caramel pigment (200 ml, C.V.
0.92) by Cl form anion exchanger (2.0 g) at 60°C. ○ : IONEX No. 2;
● : Amberlite IRA-401 (low crosslinked type); ▲ : Amberlite IRA-400
(standard crosslinked type); ■ : Amberlite IRA-900 (MR type).

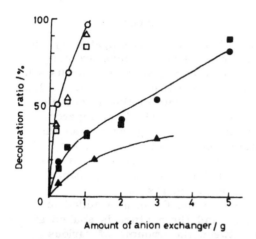

Figure 7.12 Decoloration ratio versus amount of anion exchanger. Decoloration of 50 wt% sucrose solution containing caramel pigment (25 ml, C.V. 0.92) by Cl form of anion exchanger (60°C, 6 hr). ○:IONEX No. 2; △ : IONEX No. 3; □ :IONEX No. 4; ● : IRA-401; ▲: IRA-400; ■ : IRA-900.

by the fact that the surface area ratio of the fiber to the resin $(2r_0/3r_f)$ is about 10, since the caramel pigment of a macromolecular weight cannot diffuse into the exchangers and can be adsorbed only near their surface.

It can be concluded that the fiber has a large capacity for adsorbing macromolecular ionic substances such as caramel pigment, compared with the resins.

5. APPLICATIONS

5.1 Preparation of Ultrapure Water

The ion exchange fiber has a high ion exchange rate and a large capacity for adsorbing macromolecular ionic substances such as organisms, pyrogens, and inorganic colloidals, as described above. The fiber is, therefore, applicable to the process for preparation of ultrapure water, especially as a final polisher. TORAYPURE LV-10T system for ultrapure water, shown in Fig. 13, utilizes a combination of Toray's spiral-type reverse osmosis membrane (ROMEMBRA) and special ion exchange fiber (IONEX). This system can supply ultrapure water with a resistivity of more than 18 MΩ·cm from tap water. Typical examples of running and results of water analysis are shown in Fig. 14 and Table 3, respectively.

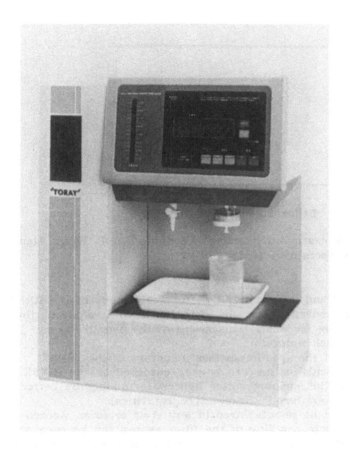

Figure 7.13 TORAYPURE LV-10T system for ultrapure water.

5.2 Water Purification in the Nuclear Energy Industry

Ion exchange resin has been utilized extensively in the nuclear energy industry. Examples are purification of light water in nuclear power reactor systems such as filters and demineralizers, for condensate between power reactors and power generators, and for treatment of water in power reactors and nuclear fuel pools. Application for purifying radioactive waste water has also been important.

The differences between IONEX and conventional resin are as follows.

1. The ion exchange rate is much faster in IONEX, and this makes it possible to reduce not only the amount of resin required, but also the size of equipment required.

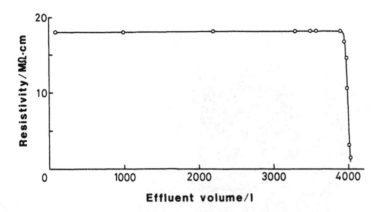

Figure 7.14   One example of running. Feedwater, 0.01 MΩ·cm (tap water); water temperature, 30-35°C.

2.   Less acid and base is required for reactivation of the IONEX than for conventional resin. With less IONEX required, the expenditure for waste management of the overall system can be much reduced.

3.   Because of the specific surface structure of the IONEX filter, colloidal oxides can be well adsorbed on the IONEX surface. This ensures better light water quality by better elimination of "crud" (radioactive impurities).

4.   IONEX can be manufactured in any style or form. Accordingly, remote handling of the filter system can be adequately designed to make operation and maintenance easier and safer.

Pretesting of IONEX for a simulated filter and demineralizer in a cold stage has been successfully carried out, and the results obtained so far are encouraging in support of advantages of IONEX. Figure 15 shows ferric oxide (particle size, 0.3 µm) adsorbed on the strong anion IONEX. A hot test using a commercial unit is planned for the near future.

5.3   Selective Removal of Heavy Metal Ions

Recently, selective chelating resins based on a styrene divinyl-benzene polymer have been studied [8-10] and used for the separation, removal, and recovery of heavy metal ions in waste fluid treatment, the atomic power industry, the food industry, and analytical chemistry. The adsorption properties of the chelating IONEX for heavy metal ions have been investigated [2].

Table 7.3   Results of Water Analysis

Item		Feedwater	Permeate
Resistivity	$(M\Omega \cdot cm)$	0.01	>18
Sodium	(ppb)	7,200	< 1
Silica	(ppb)	850	< 5
Bacteria	(No/ml)	1.5	< 0.1
Pyrogens	(ng/ml)	2.6	< 0.1
Particles	(No/ml)[a]	63,500	27
TOC	(ppb)	800	70

[a]>0.2 μm.

The correlation between the adsorption capacity of Cu^{2+} and the pH in the solution for various chelating exchangers is shown in Fig. 16. The IDA-type fiber exhibits the same pH dependence as the chelating resin (Diaion CR-10, the IDA-type resin; diameter, ca. 400 μm; water content, 1.7). The half pH, where the adsorption capacity becomes half of pH 5.2, is 1.7. On the other hand, the IDP-type fiber exhibits a different pH dependence from the IDA type; here the half pH is 2.9. These results indicate that the IDA type forms more stable complexes with Cu^{2+} than the IDP type [9]. This fact can be explained by assuming that the former coordinates to Cu^{2+} by forming five-member rings, while the latter does so by forming six-member rings.

Ion exchange between the Na form of a chelating exchanger (0.25 g) and a 0.013 M Cu^{2+} solution containing 0.5 M NaCl and a Clark Labs buffer of pH 6.0 (100 ml) has been examined. Figure 17 displays the fractional attainment of equilibrium as a function of the time. The fibers do require less than 10 min to attain 50% of equilibrium, whereas the resin requires more than 1 hr. Besides, the exchange rate becomes higher as the water content of the fibers increase, that is, as the degree of crosslinking is lowered. This means that the diffusion within the exchangers is the major rate-controlling step in the adsorption process. It can be seen from Fig. 17 that the exchange rate of the fibers is from 10 to 40 times as high as that of the resin. These results are interpreted in terms of the rate laws of ion exchange for the limiting case of ideal particle-diffusion control in a simple system [2].

A solution of 830 mg/dm^3 Cu^{2+} containing 0.5 M NaCl is passed through a column packed with a chelating exchanger at a space velocity (S.V.) of 10/hr. The breakthrough curves of the filamentary IONEX No. 4 and Diaion CR-10 columns are given in Fig. 18. The

5 μm

Figure 7.15 Ferric oxide adsorbed on strong anion IONEX.

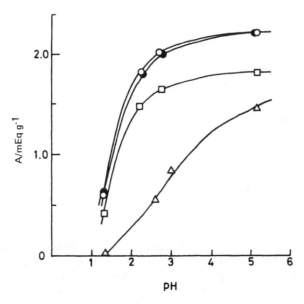

Figure 7.16 Effect of pH in solution on adsorption capacity of Cu^{2+}. Adsorption to Na form of chelating exchanger (0.25 g) was carried out with 100 ml of 0.013 M Cu^{2+} solution containing 0.5 M NaCl and Clark Labs buffer (20°C, 24 hr). ○ : IONEX No. 1; □ : IONEX No. 2; △ :IONEX No. 4; ● : Diaion CR-10.

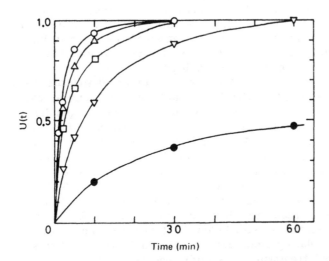

Figure 7.17 Fractional attainment of equilibrium as a function of time. Ion exchange between Na form of chelating exchanger (0.25 g) and 0.013 M Cu^{2+} solution containing 0.5 M NaCl and Clark Labs buffer of pH 6.0 (100 ml). ○ : IONEX No. 1; □ : IONEX No. 2; △ : IONEX No. 4; ▽ : IONEX No. 5; ● : Diaion CR-10.

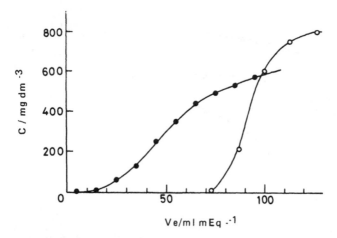

Figure 7.18 Concentration of Cu^{2+} in effluent versus effluent volume. Feed solution, 830 mg/dm³ (0.013 M) Cu^{2+} solution containing 0.5 M NaCl and Clark Labs buffer of pH 6.0; bed volume, 10 ml (1 cm × 13 cm); packed weight of chelating exchanger in Na form, 2.5 g; flow rate, 100 ml/hr (S.V. 10/hr). ○ : Filamentary IONEX No. 4; ● : Diaion CR-10.

ion leaks through the resin column in the early stage, while the fiber column represents the breakthrough curve of a roughly S form. From this curve, the band length and utilization of the ion exchange can be evaluated to be 7.8 cm and 0.71, respectively. The band length for the resin column is estimated to be more than 30 cm. These results show that the utilization of the fiber is high, since the band length is short because of its rapid adsorption proper- ties, even if the Cu^{2+} concentration is high.

Next, a 137.5 mg/dm³ Mn^{2+} solution containing 30 wt/vol% lysine is passed through the felt IONEX No. 3 and resin columns at various space velocities. Figure 19 presents the space velocity dependence of the breakthrough volume. The minimum concentrations for the resin and felt columns are 0.05 mg/dm³ and below 0.02 mg/dm³, respectively at S.V. 5/hr. In the case of the resin column, the breakthrough volume decreases rapidly as the space velocity is raised, and the ion leak takes place in the early stage at S.V. 20/hr. On the other hand, even at S.V. 80/hr, the felt column gives 72% as large a breakthrough volume as at S.V. 5/hr, and its minimum concentration is still below 0.02 mg/dm³. These results indicate that the chelating fiber can be utilized in various fields that require a rapid treatment or a high removal ratio, since it has such a high adsorption rate.

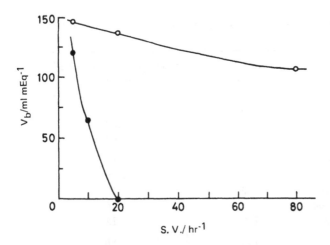

Figure 7.19 Space velocity dependence of breakthrough volume in removal of Mn^{2+} from highly concentrated lysine solution. Breakthrough point, 1 mg/dm^3 as Mn^{2+} in effluent; feed solution, 137.5 mg/dm^3 Mn^{2+} solution containing 30 wt/vol% lysine (pH 5.4). $\bigcirc$: Lysine form of felt IONEX No. 3 (1.1 g) in 5 ml (1.8 cm × 2 cm); $\bullet$: lysine form of Diaion CR-10 (2.5 g) in 10 ml (1 cm × 13 cm).

5.4 Adsorption and Immobilization of Biologically Active Proteins

Recently, the immobilization of enzymes has been intensively studied [11], and some immobilized enzymes have been used in industry. The ion exchange fiber has a large surface area and, accordingly, is expected to have a large capacity for adsorbing macromolecular proteins. The fiber has been studied for its ability to adsorb and immobilize biologically active proteins [12]. A strong cation IONEX can effectively adsorb hemoglobin. Albumin, invertase, and glucose isomerase are readily adsorbed to a strong anion IONEX. The adsorption of these proteins to IONEX takes place through an electrostatic force.

The results of albumin adsorption to anion exchangers are summarized in Table 4. The adsorption capacity for albumin (isoelectric point, pH 4.7; molecular weight, ca. 70,000) becomes exceedingly large (ca. 300 mg/g) with an increase in the water content of IONEX. Apparently, water in the polystyrene network of IONEX plays an important role in the macromolecular protein's ability to have access to the ion exchange site. In fact, IONEX comprises a polystyrene network crosslinked between the phenyl groups by methylene linkages. The lattice of the network becomes wider as the degree of crosslinking is decreased; that is, the water content of the fiber increases,

Table 7.4 Results of Albumin Adsorption to Anion Exchangers in Cl Form

Anion exchanger		Water content	Adsorption capacity[a] (mg/g)
IONEX		1.5	30
		2.5	200
		3.5	300
Amberlite	IRA-400	0.8	< 10
	IRA-401	1.6	< 10
	IRA-900	1.6	13
	IRA-904	1.4	40
	IRA-938	2.7	12
Powdex PAO		1.4	< 10

[a]0.1% albumin solution containing 0.05 M phosphate buffer (pH 7.0), 30°C, 3 hr.

and thus, a macromolecule can diffuse into the network and more easily interact with the ion exchange sites.

Table 5 shows the results of invertase immobilization to anion IONEX in Cl form by the batch method. The adsorption capacity for invertase (isoelectric point, pH 3.8; molecular weight, 270,000) is greater when the water content of the ion exchanger is higher. This fact is attributable to the difference in the crosslinked structure described above. The yield of the activity is enhanced from ca. 40% to ca. 50% when the water content of the fiber is increased from 1.5 to 2.5. A continuous inversion of sucrose has also been investigated by using the immobilized invertase [12]. From these results, this fibrous ion exchanger with a high water content is considered to be an excellent material for the adsorption and immobilization of proteins.

5.5 Adsorption of Microorganism Cells

Recently, immobilization of microorganism cells as well as enzymes has been intensively studied from a viewpoint of their use as bioreactors in industry [13]. Immobilization of microorganism cells by adsorption on the ion exchange fiber has been studied [14].

The microorganism cells, such as yeasts, bacteria, and actinomycetes, are well adsorbed on anion IONEX. They are slightly adsorbed on a macroreticular-type anion exchange resin (Amberlite IRA-938).

Table 7.5   Results of Invertase Immobilization to Anion IONEX
in Cl Form by Batch Method

Water content	Amount of IONEX (mg)	Adsorbed[a] enzyme (A) (U)	Observed activity (B) (U)	Yield of activity (B/A) (%)
1.5	250	610	230	38
	500	1170	460	39
	1000	2125	860	40
2.5	50	625	350	56
	150	1940	1025	53
	500	2125	1140	54

[a]Enzyme added: 2125 U, 12.5 mg (50 ml of 0.01 M phosphate buffer,
pH 5.0), 20°C, 2 hr.

These results suggest that the outer walls of the microorganism
cells are negatively charged and are adsorbed on anion exchangers
by electrostatic force. Figure 20 shows the relation between the
cell adsorption capacity and the water content of anion IONEX.
The adsorption capacity is dependent on the water content of the
fiber, but independent of the kind of anion exchange group. Upon
increasing the water content of the fiber, the adsorption capacity
is considered to increase rapidly, since the adherent force of the
cells to the fiber is enhanced by the affinity for water and this
inhibits the repulsion between the cells.

Figures 21 and 22 show G-actinomycetes and H-yeasts adsorbed
on the strong anion IONEX. The adsorption and desorption behavior
of the cells is different between actinomycetes and yeasts, and
also between strong and weak anion exchange fibers. The enzyme
activities of the immobilized actinomycetes containing glucose isomerase
and yeasts containing L-aminolactam hydrolase are ca. 70% and 60%
of those of the native cells, respectively. The stability of the immobilized yeasts in the hydrolysis of DL-cyclic lysine anhydride
has also been investigated [14].

5.6 Solid Acid-Base Catalysts

Polystyrene-based ion exchange resins have been intensively studied
as solid acid-base catalysts in a number of organic reactions [15].
The cation IONEX has also been investigated as a solid acid catalyst
in some organic reactions [16]. The reactions of sucrose inversion
and methyl acetate hydrolysis are carried out by pumping a 50 wt%

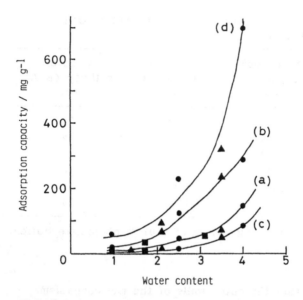

Figure 7.20 Cell adsorption capacity versus water content of anion IONEX: (a) G-actinomycetes, (b) H-yeasts, (c) R-bacteria, (d) P-yeasts. ● : Trimethylammonium group; ■ : dimethylamino group; ▲ : amino group.

sucrose aqueous solution at 60°C and a 3.2 M methyl acetate aqueous solution at 30°C, respectively, through a reactor packed with the fibrous cation exchange catalysts. The reactions are also carried out by using commercially available cation exchange resins (Amberlite IR-120B and IR-200) with radii of ca. 250 μm.

The results of sucrose inversion by various acid catalysts are shown in Fig. 23. It is found that the height of the fixed bed and the crosslinking degree of the fiber do not affect the rate constant. The catalytic activity of the knitted IONEX is 30% lower than that of the filamentary IONEX. This result may be interpreted by assuming that the effective surface area per unit of weight decreases or an irregular flow takes place because of the knitted structure. The apparent rate constant of the filamentary IONEX is about 16 times as large as that of the gel-type IR-120B and macroreticular-type IR-200 resins. These results can be interpreted in terms of the solid catalysis theory [16]. These ion exchange fibers have not only much higher catalytic activities than ordinary ion exchange resins when the reactant is a large molecule such as sucrose, but also the advantage of easier catalytic treatment than a fine or crushed resin.

Figure 7.21 Adsorption state of G-actinomycetes on strong anion IONEX of water content 2.5.

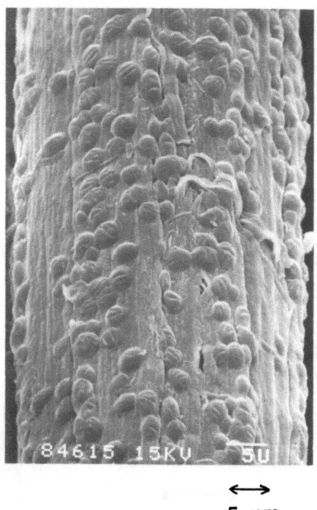

5 μm

Figure 7.22 Adsorption state of H-yeasts on strong anion IONEX of water content 2.5.

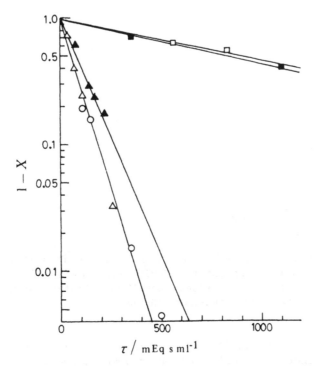

Figure 7.23 Sucrose ratio versus contact time in sucrose inversion. Reaction temperature, 60°C; concentration of sucrose solution, 50 wt%. ○ : Filamentary IONEX (water content 2.0); △ : filamentary IONEX (water content 1.6); ▲ : knitted IONEX (water content 1.6); ☐: Amberlite IR-120B; ■ : Amberlite IR-200.

Figure 24 shows the results of methyl acetate hydrolysis by various acid catalysts. No large difference in the apparent rate constants is observed between the fiber and IR-120B. On the other hand, the apparent rate constant for IR-200 is only one-third as large as that for IR-120B. When the reactant is a small molecule such as methyl acetate, these ion exchange fibers show as high catalytic activities as ordinary ion exchange resins.

The fibrous catalyst is expected to be applicable to new reactors such as the reaction distiller of methyl acetate hydrolysis, since it can be used in various forms.

5.7 Phase Transfer Catalysts

Fiber-supported quaternary phosphonium ion and crown ether have been investigated as phase transfer catalysts for numerous organic

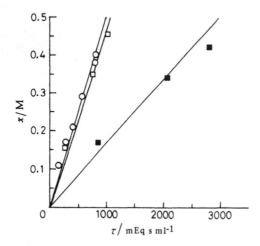

Figure 7.24 Acetic acid concentration versus contact time in methyl acetate hydrolysis. Reaction temperature, 30°C; concentration of methyl acetate solution, 3.2 M. ○ : Filamentary IONEX (water content 2.0); □ : Amberlite IR-120B; ■ : Amberlite IR-200.

reactions by Tomoi and Kakiuchi [3]. They have reported the activity of the fibrous (cut fiber form) and spherical triphase catalysts for the reaction of 1-bromooctane with sodium cyanide or potassium iodide at 90°C. The fibrous phosphonium catalyst with a loading of near to 1 mEq/g shows a large activity, which is comparable to that of 2-10 mol% crosslinked, 74-149 μm (100-200 mesh) bead catalysts. The activity of the fibrous crown catalyst is somewhat less than that of the corresponding bead catalysts. The fibrous catalysts are separated from the reaction system by filtration more easily than the bead catalysts are, since the fibrous catalysts are 1 mm long.

The polystyrene-polypropyrene composite fibers in various forms as well as polystyrene beads can also be used as supports for other catalysts or reagents. The fiber-supported reagents or catalysts are easy to treat compared with ordinary bead reagents or catalysts, and can be applied to practical organic reactions, especially large-scale productions or continuous reactors (fixed or fluidized bed).

5.8 Removal of Mutagenic Components in Cigarette Smoke

Kier et al. have investigated the mutagenic activity of cigarette smoke condensate, and they have revealed that a major portion of the mutagenic activity of the whole condensate is in basic fractions [17]. Therefore, it is expected that the cation exchange fiber (IONEX) will be useful to remove the basic mutagens in cigarette smoke.

In fact, we have confirmed in a model smoking test that insertion of a small amount of the cation exchange fiber into a conventional cellulose acetate cigarette filter results in a remarkable decrease not only in the quantity of smoke condensate, but also in mutagenic activity of the whole condensate [18,19].

Furthermore, in an actual smoking test, it has been confirmed that the urine from smokers of cigarettes containing a small amount of the cation exchange fiber exhibits much weaker mutagenic activity compared with that from regular cigarette smokers.

5.9 Adsorbent for Blood

5.9.1 Adsorption of Bilirubin

Bilirubin is one of the metabolic products of hemoglobin and harmful to the brain. In the human body, bilirubin concentration in blood is normally maintained below 1 mg/dl. But when the function of the liver lowers or when cholangitis is closed, the concentration increases to more than 10 mg/dl, and sometimes to 30 mg/dl. A concentration greater than 10 mg/dl is considered to be harmful to the brain. To prevent brain attack, it is necessary to remove bilirubin from the blood. Though bilirubin is a low molecular weight compound of 584 daltons, it cannot be removed by dialysis because it exists in conjugated form with albumin in blood. Therefore, many studies of bilirubin removal have been done, but no satisfactory adsorbent has yet been made whose adsorptive capacity is sufficiently large.

We calculated the required capacity under the following six assumptions:

1. Body weight of the patient is 60 kg.
2. Hematocrit of the blood is 0.35.
3. Bilirubin concentration in the plasma of the patient is 30 mg/dl.
4. Safety region of bilirubin concentration is below 10 mg/dl in plasma.
5. Total amount of bilirubin in the body is three times that in the plasma.
6. Maximum amount of an adsorbent it is possible to use at a time is 100 g.

The result was a capacity of more than 18 mg per 1 g of adsorbent at the bilirubin concentration of 10 mg/dl.

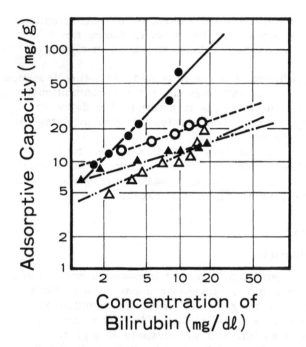

Figure 7.25 Adsorption isotherms of bilirubin in bovine serum for 3 hr at 37°C for Fiber A (○), Fiber B (●), anion exchange resin Amberlite IRA-938 (▲), and powdery activated charcoal (△). Each adsorbent is fully heparinized with 0.1 mg/ml of heparin aqueous solution prior to the experiment.

Bilirubin is expected to be adsorbed by powdery activated charcoal or anion exchange resin, because it is a hydrophobic dicarboxylic acid. However, according to our estimation by adsorption isotherm, the capacity is not so large, as shown in Fig. 25. For example, it was 12 mg/g for powdery activated charcoal and 13 mg/g for strong anion exchange resin (Amberlite IRA-938) at a bilirubin concentration of 10 mg/dl. Neither of them exceeds the above-mentioned criterion of 18 mg/g. On the other hand, it is 19 mg/g for the fiber containing dimethylaminomethyl group on the aromatic nucleus (Fiber A) and 50 mg/g for that of IV (Fiber B).

$$-CH_2CH-$$

$$CH_2NHCOCH_2\overset{\oplus}{N}(CH_2CH_2CH_3)_3 \ \overset{\ominus}{Cl}$$

(IV)

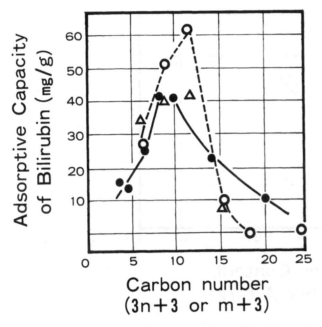

Figure 7.26 Effect of the number of carbon atoms contained in alkyl substituents on adsorptive capacity of bilirubin in polystyrene-derivative fibers. $\bigcirc$, $\triangle$:

$$-CH_2CH-$$
$$\bigcirc\!\!\!\!-CH_2NHCOCH_2\overset{\oplus}{N}\left\{(CH_2)_nCH_3\right\}_3\overset{\ominus}{Cl};$$

$\bullet$:

$$-CH_2CH-$$
$$\bigcirc\!\!\!\!-CH_2NHCOCH_2\overset{CH_3}{\underset{\overset{|}{CH_3}}{\overset{\oplus|}{N}}}-(CH_2)_mCH_3;\quad \triangle : \text{nonheparinized IONEX;}$$

$\bigcirc$, $\bullet$: heparinized IONEX.

     To investigate whether functional group IV is specific for adsorbing bilirubin or not, the fibers containing functional group II of various alkyl groups were prepared and their adsorptive capacities were measured. As shown in Fig. 26, when the sum of the carbon numbers contained in their three alkyl groups (R_1, R_2, and R_3) is 6 to 12, the fiber gives the largest ability. And as shown in Figs. 27–29, the adsorptive capacity increases in proportion to the water content in Fibers A, B, and C containing the functional group V.

$$-CH_2CH-$$
$$\bigcirc\!\!\!\!-CH_2NHCOCH_2\overset{\oplus}{N}(CH_2CH_2CH_2CH_3)_3\overset{\ominus}{Cl}$$

$$(V)$$

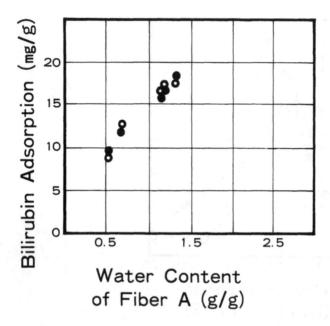

Figure 7.27 Effect of water content on adsorptive capacity of bili-rubin at the concentration of 10 mg/dl in Fiber A, heparinized (●) and nonheparinized (○).

5.9.2 Adsorption of Heparin

In extracorporeal therapy such as dialysis, an anticoagulant is requisite to prevent blood coagulation. Heparin is used in almost all these cases. When the amount of heparin is too little, blood coagulation occurs; and, on the contrary, when too much heparin is used bleeding from mucous membrane happens. Therefore, the control of heparin amount is very important. So the adsorbent that tends to adsorb a lot of heparin is hard to deal with.

As shown in Fig. 30, the shorter the length of alkyl group in the fiber, the more the fiber adsorbs heparin about analogs of Fiber B. From the results of Figs. 26 and 30, it may be concluded that functional groups IV and V are the best. And the increase in the adsorptive capacity of heparin is linear with increase in water content of the fiber as that of bilirubin among fibers having the same functional group. Consequently, there is the same relation between adsorptive capacity of heparin and that of bilirubin as shown in Fig. 31.

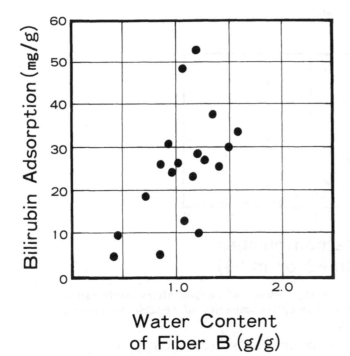

Figure 7.28 Effect of water content on adsorptive capacity of bilirubin at the concentration of 10 mg/dl in Fiber B, heparinized.

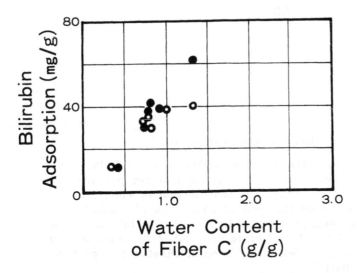

Figure 7.29 Effect of water content on adsorptive capacity of bilirubin at the concentration of 10 mg/dl in Fiber C, heparinized (●) and nonheparinized (○).

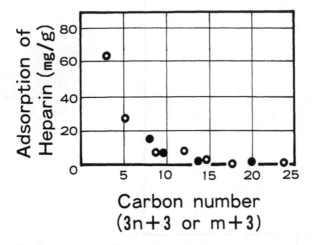

Figure 7.30 Effect of the number of carbon atoms contained in alkyl substituents on adsorptive capacity of heparin in Fiber B analogs.

5.9.3 Adsorption of Other Serum Components

Figure 32 shows adsorption spectra of Fibers A and B for serum components contained in blood from human jaundice patients. The spectrum of Fiber C is almost the same as that of Fiber B. From the results, these fibers turn out to adsorb harmful substances, such as bile acids and lipid peroxides, as well as bilirubin, effectively. However, Fiber A, differing from Fiber B, tends to adsorb protein components. So, when the harmful substance is protein, Fiber A is very useful.

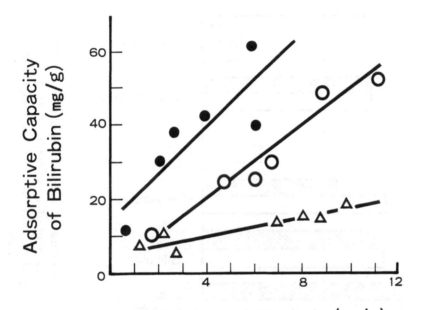

Figure 7.31   Adsorptive capacity of bilirubin at 10 mg/ml versus adsorption of heparin by fibers having following functional groups:

$\bigcirc$: $\begin{matrix} -CH_2CH- \\ \bigcirc \end{matrix}$ $\!\!-CH_2NHCOCH_2\overset{\oplus}{N}(CH_2CH_2CH_3)_3\overset{\ominus}{Cl}$; $\triangle$: $\begin{matrix} -CH_2CH- \\ \bigcirc \end{matrix}$ $\!\!-CH_2N(CH_3)_2$;

$\bullet$: $\begin{matrix} -CH_2CH- \\ \bigcirc \end{matrix}$ $\!\!-CH_2NHCOCH_2\overset{\oplus}{N}(CH_2CH_2CH_2CH_3)_3\overset{\ominus}{Cl}$.

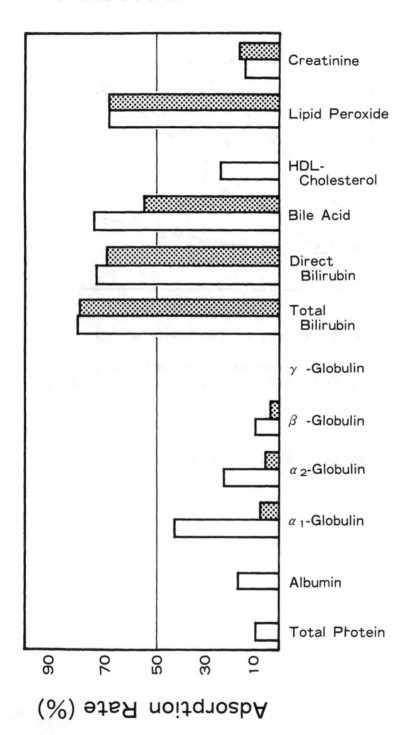

Figure 7.32 Adsorption spectra of Fibers A (□) and B (■) in the serum of the jaundiced patient. Intact serum contained 4.50 g/dl of total protein, 2.65 g/dl of albumin, 0.31 g/dl of α_1-globulin, 0.12 g/dl of α_2-globulin, 0.27 g/dl of β-globulin, 1.06 g/dl of γ-globulin, 17.9 mg/dl of total bilirubin, 4.50 mg/dl of direct bilirubin, 9.9 nmol/dl of bile acids, 8 mg/dl of HDL-cholesterol, 11.4 nmol/dl of lipid peroxides, and 5.2 mg/dl of creatinine.

6. CONCLUSION

Polystyrene-based functional fibers that are excellent in chemical stability and mechanical strength have been prepared by using the islands-in-a-sea type of composite fiber as the starting material. The fiber has a large surface area per unit of weight and can be used in various forms.

The resulting ion exchange fiber has two fundamental characteristics: The ion exchange rate for metal ions is extremely high, and the capacity for adsorbing the macromolecular ionic or polar substances is exceedingly large compared with ordinary ion exchange resins.

As the ion exchange rate is extremely high, the ion exchange fiber is expected to be applicable to the processes of preparation of ultrapure water, treatment of condensed water for an atomic power plant, removal of harmful ions, and recovery of useful ions: It is suitably applicable to treatment for a dilute solution.

Since the capacity adsorbing macromolecular ionic substances is exceedingly large, the fiber has been studied as a material for adsorbing or immobilizing biologically active substances, such as enzymes, proteins, and microorganism cells, and for adsorbing coloring matters or colloidal substances.

The fiber has also been studied as solid acid-base catalysts in organic reactions, and is expected to be applicable to new reactors such as the reaction distiller.

The anion exchange fibers with special functional groups have been investigated as adsorbents for harmful serum components, and is expected to be applicable to the removal of bilirubin in blood from human jaundice patients.

REFERENCES

1. T. Yoshioka and M. Shimamura, *Bull. Chem. Soc. Jap.*, *56*, 3726 (1983).
2. T. Yoshioka, *Bull. Chem. Soc. Jap.*, *58*, 2618 (1985).
3. M. Tomoi and H. Kakiuchi, *Makromol. Chem.*, *Rapid Commun.*, *5*, 685 (1984).
4. K. Teramoto, M. Shimamura, and T. Yoshioka, Japanese Disclosed Patent 11,210, 1982.
5. K. Teramoto, T. Miyoshi, and M. Shimamura, Japanese Disclosed Patent 134,993, 1978.
6. K. Teramoto, T. Miyoshi, and M. Shimamura, Japanese Disclosed Patent 110,274, 1979.
7. K. Teramoto, M. Sakamoto, and M. Shimamura, Japanese Disclosed Patent 12,008, 1982.

8. F. Helfferich, *Ion Exchange*, McGraw-Hill, New York, 1962, pp. 45-47, 259-264.
9. G. Kühn and W. Hoyer, *Makromol. Chem.*, *108*, 84 (1967).
10. E. Blasius and B. Brozio, *Chelates in Analytical Chemistry* (H. A. Plaschka and A. J. Barnard, Jr., Eds.), Marcel Dekker, New York, 1967, p. 49.
11. L. Goldstein and G. Maneche, *Immobilized Enzyme Principles*, Academic Press, New York, 1976, pp. 25-30.
12. T. Yoshioka and M. Shimamura, *Bull. Chem. Soc. Jap.*, *59*, 399 (1986).
13. I. Takada, T. Tosa, and I. Chibata, *J. Solid-Phase Biochem.*, *2*, 225 (1977).
14. T. Yoshioka and M. Shimamura, *Bull. Chem. Soc. Jap.*, *59*, 77 (1986).
15. F. X. McGarvey and R. Kunin, *Ion Exchange Technology* (F. C. Nachod and J. Schubert, Eds.), Academic Press, New York, 1956, p. 272.
16. T. Yoshioka and M. Shimamura, *Bull. Chem. Soc. Jap.*, *57*, 334 (1984).
17. L. D. Kier, E. Yamasaki, and B. N. Ames, *Proc. Natl. Acad. Sci. USA*, *71*, 4159 (1974).
18. S. Yoshikawa, T. Yoshioka, and M. Shimamura, U.S. Patent 4,700,723, 1987.
19. M. Noyori, S. Yoshikawa and M. Shimamura, Japanese Disclosed Patent 183,022, 1985.

8

BIOABSORBABLE FIBERS FOR MEDICAL USE

YOSHITO IKADA / Kyoto University, Kyoto, Japan

1. INTRODUCTION

Within the two last decades biomedical materials have increasingly
attracted much attention of material scientists and medical doctors.
This is primarily because recent rapid progress in the material
science has made possible the use of synthetic materials in the
medical field, which in turn has necessitated introduction of advanced
new technologies for its further development. It is said that one
of the most striking events in clinical medicine in the middle of
this century is the advent of biomaterials, which comprise the essential parts of artificial organs and medical devices.

Biomaterials are composed of metals, ceramics, polymers, and their composites. Each has advantages and disadvantages. For instance, the polymeric materials are diversed in modulus, from flexible elastomer to strong fiber, and are readily processed into products of various shapes. However, plastics do not have such high strengths as ceramics and metals and are lower in chemical stability than ceramics.

We can no longer imagine modern life without fibers. So is it also in the surgery, where incision, excision, and closure of the body are the basic unit operations. Suturing with fibrous materials is indispensable for tissue closure. It is not impossible to make fibers from metals and ceramics, but pliable, well-handling fibers with wide applications can be fabricated only from polymers. Also, the fibers used in surgery are produced overwhelmingly from polymers.

Fiber products have been commonly used in medicine. Well-known examples are (a) hollow fibers for artificial kidneys and lungs, (b) woven and knitted tubes from polyester for vascular grafts, and (c) threads for suture. Cloths for patient bedsheets and operating gowns for surgeons are also made from fiber, but they are generally not included in biomaterial, which is defined as the material to be used in direct contact either with body fluids such as blood or with living tissues.

From the point of view of bioabsorbability, biomaterial can be divided into two groups, bioabsorbable and nonbioabsorbable. The former has been used as a temporary assist material for scaffolding, supporting, covering, or filling damaged tissue until it heals. The most typical application is the bioabsorbable suture. Such a bioabsorbable material is likely to become increasingly important among biomaterials, because it has the unique advantage that it does not remain in body as a foreign body until the death of body, but disappears from the body when it has fulfilled its function as a temporary assist.

This chapter describes only bioabsorbable fiber, an attractive and promising biomaterial, since nonbioabsorbable fibers have been so widely applied to medicine that they are beyond the scope of this chapter.

2. REQUIREMENTS FOR MEDICAL USES

When a material is applied in medicine, special properties are required for the material in addition to those common to industrial materials, such as mechanical strength and productivity. It seems, therefore, better to summarize briefly the requisites demanded of the material to be used for medical purposes.

Table 8.1   Requirements for Polymer as Biomedical Materials

Properties	Description	Degree of importance
1. Nontoxicity	Nonpyrogenicity, nonallergenic response, noncarcinogenesis	Essential
2. Sterilizability	Radiation, ethylene oxide gas, dry heating, autoclave	Essential
3. Sufficient mechanical properties	Strength, elasticity, durability	Essential
4. Biocompatibility	Bioinert, bioactive	Desirable

The properties necessary or desired for biomaterials are listed in Table 1. The most important attribute that distinguishes a biomaterial from an industrial one is toxicity of the material to body. Toxicity results in various adverse tissue reactions. When a material implanted in the body elicits body temperature rise, chronic tissue inflammation, red blood cell lysis, allergic reaction, carcinoma, or deformity, the implanted material is regarded as toxic. Whether the toxicity is mild or severe, acute or chronic, depends largely on the material's history and purification. Since a material contaminated with bacteria—that is, infected—exhibits toxicity, sterilization of the material is absolutely needed. Therefore, the material must be capable of being sterilized without detriment. Commonly used sterilization methods involve gamma-ray irradiation, ethylene oxide gas treatment, heating in the dry state, or steaming in autoclave. Radiation sterilization may become the most common method in the future, because ethylene oxide is somewhat toxic and costly. Many of the polymeric biomaterials are not as heat resistant, as tolerable to sterilization at high temperatures. Hydrolyzable materials should not be exposed to steam.

Some materials exhibit toxicity even though carefully sterilized. In such cases the toxicity is caused mostly by low molecular weight compounds that have leached out from the biomaterial. Among the leachable compounds are polymerization or crosslinking catalyst fragments, monomers that remain unpolymerized, and additives such as antioxidants, ultraviolet absorbents, pigments, and fillers. Materials of medical grade are generally fabricated by careful and rigorous purification of raw materials to minimize the content of additives.

Other possible sources of leachable compounds are biodegradation products or partial dissolution of the biomaterial itself. Toxicity of metallic biomaterials has been often reported to result from their dissolved metal ions. In the case of polymeric materials, silicone, polytetrafluoroethylene, and polyolefins have preferably been utilized, especially in long-term implants, chiefly because of their high chemical resistance. Safety of poly(vinyl chloride) plasticized with 2-ethylhexyl phthalate has been always an object of controversy.

All of the bioabsorbable materials undergo biodegradation to yield low-molecular-weight biodegradation products that might potentially give adverse effects to cells by interacting with the cell surface or penetrating the cell membrane into the interior. Therefore, selection of bioabsorbable materials must be made with much more caution than for nonbioabsorbables. A great challenge is the synthesis of such a polymer that has excellent mechanical properties with an adequate biodegradation rate, is capable of being sterilized with gamma-radiation, and generates nontoxic biodegradation products.

The biocompatibility described in Table 1 means more than non-toxicity and includes bioinertness and bioadhesiveness. A bioinert material does not elicit severe foreign body reactions such as activation of the immune system or blood coagulation, leading to encapsulation and thrombus formation. So far, no material has been reported that is entirely bioinert. Also, no bioadhesive material that is able to bind firmly to surrounding tissue has yet been developed except where tissue-like materials are used. Such biocompatibility is mostly required for biomaterials used in long-term implantation and is not very crucial for bioabsorbable materials that remain in the body for relatively short periods of time.

3. BIODEGRADATION AND BIOABSORPTION

3.1 Terminology

The definition of biodegradation is still vague and controversial. Indeed, a variety of terms associated with biodegradation have been used, including degradation, bioabsorption (absorption), bio-erosion (erosion), and resorption. Some people use the prefix "bio-" when a material is degraded only in a biological environment, not in a nonbiological environment. Degradation of polypeptides normally needs enzyme and does not take place in aqueous media without enzymes, whereas aliphatic polyesters undergo significant degradation in the presence of water, practically regardless of the presence of hydrolytic enzymes. Thus, strictly speaking, polypeptides are biodegradable, while aliphatic polyesters are not biodegradable. In the present chapter, the term biodegradation is used in a wide sense, that is, for any degradation occurring in biological environments, even if an enzyme is not necessary. Thus, aliphatic polyesters are biodegradable under our definition.

Microscopic degradation of polymeric materials results from scission of main chains of the polymer molecules. Two modes of chain scission are known; one is random chain scission and the other is zipperlike scission from the end of the polymer chain. The mass of the polymer material, as well as the molecular weight, decreases linearly with the zipperlike scission, and the degraded polymer is gradually bioabsorbed into the surroundings. On the contrary, random scission is not accompanied by a significant loss of mass, especially in the initial stage, but results in remarkable reduction in molecular weight and mechanical strength. If a detectable loss of mass takes place at all, even in a much later stage than the molecular weight reduction, we will regard the polymer as bioabsorbable. Generally, nylon is not classified as a bioabsorbable polymer, because the rate of bioabsorption is too low compared with the rate of chain scission.

Macroscopically, there are at least two types of bioabsorption or resorption. One is the bioabsorption that starts from the outermost surface region of the material and goes to the inside in much the same manner as peeling a fruit. In this case the diameter of the fiber decreases with the biodegradation time. The other type of biodegradation takes place almost homogeneously throughout the cross section of the fiber from the beginning of the biodegradation. Such a case is seen when water molecules readily penetrate into the inside of the polymer material and an enzyme is not required. The polymer that disappears according to the former mode of bioabsorption is called bioerosible.

A bioabsorbable polymer is finally degraded into water soluble, low molecular weight compounds such as the repeating unit, which are then absorbed by the surrounding body fluid. However, there is an exceptional group of polymeric materials that do not yield low molecular weight compounds upon biodegradation but are bioabsorbable. These materials are composed of crosslinked polymer chains. If the individual chain is water soluble and the crosslinking bonds are biodegraded, the polymer material disappears and seems to have been bioabsorbed as a result of biodegradation.

3.2 Bioabsorbable Polymers

As mentioned above, we use the terms biodegradable and bioabsorbable without clearly distinguishing them.

All of the known bioabsorbable polymers are degraded through hydrolysis of bonds in the main chain. Water molecules are therefore absolutely required, but enzymes are not always necessary. However, when the polymer material is degraded into fragments smaller than approximately 10 μm, phagocytic cells such as macrophages may engulf the small fragments to accelerate the bioabsorption.

Table 8.2 Naturally Occurring Biodegradable Polymers

Hydrolyzable unit		Polymer
Name	Structure	
Peptide	$-\overset{\displaystyle \ \ \ }{\underset{\displaystyle H}{N}}-\overset{\displaystyle O}{\overset{\displaystyle \|}{C}}-$	Collagen, fibrin, albumin, gelatin
Hemiacetal	-O-	Starch, hyaluronic acid, chitin
Ester	$-\overset{\displaystyle }{\underset{\displaystyle O}{\overset{\displaystyle \|}{C}}}-O-$	Poly-β-hydroxybutyrate (PHB), poly(malic acid)
Phosphate	$-\overset{\displaystyle O}{\underset{\displaystyle O}{\overset{\displaystyle \|}{\underset{\displaystyle \|}{P}}}}-O-$	RNA, DNA

Table 8.3 Synthetic Biodegradable Polymers

Hydrolyzable unit		Polymer
Name	Structure	
Ester	$-\overset{\displaystyle }{\underset{\displaystyle O}{\overset{\displaystyle \|}{C}}}-O-$	Polylactides, polylactones, aliphatic polyesters
Anhydride	$-\overset{\displaystyle }{\underset{\displaystyle O}{\overset{\displaystyle \|}{C}}}-O-\overset{\displaystyle }{\underset{\displaystyle O}{\overset{\displaystyle \|}{C}}}-$	Polyanhydrides
Carbonate	$-O-\overset{\displaystyle }{\underset{\displaystyle O}{\overset{\displaystyle \|}{C}}}-O-$	Polycarbonates
Orthoester	$\overset{\displaystyle -O \quad \ O-}{\underset{\displaystyle -O \quad \ C-}{C}}$	Polyorthoesters
Carbon-carbon	$-CH_2-\overset{\displaystyle CN}{\underset{\displaystyle COOR}{C}}-$	Poly-α-cyanoacrylates
Nitrogen-phosphorus	$-N=\overset{\displaystyle \|}{\underset{\displaystyle \|}{P}}-$	Polyphosphazenes

The bioabsorbable materials include a variety of natural and synthetic polymers as exemplified in Tables 2 and 3. To be applied as fiber, the bioabsorbable polymer should be spinnable and have relatively high tensile strength and hence must be crystalline. The requirement of crystalizability imposes great restrictions to the choice of polymer as bioabsorbable fiber. This is the reason why only a few polymers listed in Tables 2 and 3 have been used clinically or are being developed as bioabsorbable fiber. They are native and regenerated collagens, synthetic polypeptides, oxidized cellulose, chitin, poly-β-hydroxybutyrate, and polylactides [poly(glycolic acid), poly(L-lactic acid), and their copolymers].

4. SURGICAL APPLICATIONS OF BIOABSORBABLE FIBERS

Bioabsorbable fibrous materials have a wide variety of medical applications in the forms of various shapes including continuous threads, knitted and woven fabrics, nonwoven felts, and meshes.

4.1 Threads

The most common implant in surgical practice is sutures, which are sterilized threads or yarns used to close wounds until they heal adequately for self-support. The healing rate of various tissues differs as shown in Fig. 1 [1]. It is estimated that more than 10 billion sutures are implanted annually into patients throughout the world. There are two major classes of sutures: absorbable (or biodegradable), used mainly to close internal wounds; and nonabsorable (or nonbiodegradable), used mainly for exposed or cutaneous wound closure. The absorbable suture is degraded in body tissues to soluble products and disappears from the implant site, usually within 2 to 6 months. Sutures are fabricated as monofilaments or multifilaments; the latter are generally braided but sometimes twisted or spun, and may be coated with wax, silicone, fluorocarbons, or other polymers to decrease capillarity and improve handling or functional properties.

The commonly used suture materials are listed in Table 4 [1].

4.1.1 Naturally Occurring Absorbable Sutures

In early Roman days, Galen described gut ligature as a substitute for Celtic linen [2]. Since then, no suture material has been so prevalent as catgut, which is derived from the small intestines of animals, usually the outer serosal layer in cattle or the submucosal layer in sheep. The derivation of the term "catgut" is obscure.

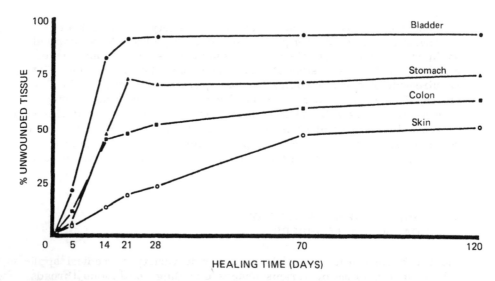

Figure 8.1 Wound strength regain in differing tissues.

Until the recent discovery of synthetic absorbable suture materials, catgut was the only practical absorbable suture material available.

For fabrication of surgical catgut, the intestinal tracts of sheep and cattle are slit lengthwise after removal of food residues and superfluous soft tissues by mechanical and chemical stripping. The ribbons are treated with a chromic salt solution to prolong the in vivo strength retention, and then an appropriate number of ribbons are twisted into bonded strands, depending on the size desired. The chromic salts may leach from the treated sutures into the surrounding aqueous environment as exemplified in Table 5 [3].

As this complex natural material has imperfections in reproducible fabrication and reliability, several approaches have been made to achieve a more uniform material by preparing regenerated collagen strands by wet spinning of dispersions of purified collagen obtained from animals. However, the regenerated collagen material has found only limited acceptance as a catgut.

It has been pointed out that catgut elicits relatively severe tissue reactions. Typical results of histochemical reactions are shown in Fig. 2 [4]. The bioabsorption of catgut is dependent on the action of proteolytic enzymes and hence its rate differs from one tissue site to another.

Catgut is packaged in solutions of aqueous alcohol, since it becomes too stiff to be handled when dry. To improve the handling qualities and remove the need to package the chromic catgut in

Table 8.4 Absorbable and Nonabsorbable Sutures

	Absorbable
Catgut	Biological origin from sheep or ox intestine; twisted monofilament; undyed and uncoated; absorbed in approximately 90 days.
Collagen	Biological origin from ox Tendo Achilles, rolled monofilament; undyed and uncoated; absorbed in approximately 90 days.
Glycolide homopolymer	Synthetic homopolymer; braided multifilament; dyed or undyed; coated or uncoated. Trade name Dexon (Davis & Geck) (PGA = polyglycolic acid); absorbed in approximately 90 days.
Glycolide and lactide co-polymer	Synthetic copolymer; braided multifilament; dyed or undyed; coated or uncoated. Trade name Vicryl (Ethicon). Ratio of glycolide to lactide is 90/10 (polyglactin 9/10). Absorbed in approximately 90 days.
Polydioxanone	Synthetic copolymer; monofilament; dyed or undyed. Trade name PDS (Ethicon) (polydioxanone suture). Absorbed in approximately 180 days).
	Nonabsorbable
Silk	Biological origin from silkworm; braided multifilament; dyed or undyed.
Linen	Biological origin from flax plant; twisted multifilament; dyed or undyed.
Cotton	Biological origin from cotton seed plant, twisted multifilament; dyed or undyed.
Polyester	Synthetic; monofilament or multifilament; dyed or undyed; coated or uncoated.
Polyamide	Synthetic; monofilament or multifilament; dyed or undyed; generic name nylon 6 or nylon 6,6.
Polypropylene	Synthetic; monofilament; dyed or undyed.
Steel	Synthetic; monofilament or multifilament.

Table 8.5   Leachable Chromium from Chromicized Catgut Sutures

Sample	USP[a] (μg g^{-1})	APHA[b](μg g^{-1})
1	75	1540, 1280, 1330
2	4	0
3	44	250, 220, 216
4	2	28
5	1	0
6	4	2
7	1	31
8	<1	1

[a]The USP specification is not more than 1 μg ml^{-1} solution or 100 μg g^{-1} suture.
[b]American Public Health Association.

alcohol, a new product has recently been introduced that incorporates glycerin in the processing of the chromic catgut material. According to the study by Stone et al., no difference was found in the degree of inflammatory response between the chromic catgut and the chromic catgut treated with glycerin [5]. Scanning electron micrographs showed a more uniform surface for the glycerin-treated chromic catgut.

Recently, a new catgut suture with rapid strength loss property has been developed by Ethicon, Inc. [6].

4.1.2 Synthetic Absorbable Sutures

As described above, the natural absorbable suture materials have several drawbacks—for instance, poor strength retention, the unwanted side effects of collagen, and the limited commercial availability of catgut.

As listed in Table 4, three kinds of synthetic, absorbable polyester sutures are now marketed: glycolide homopolymer (Dexon, Davis and Geck Division of American Cyanamide Company; and Medifit, Japan Medical Supply Co., Japan); glycolide/lactide (90/10) copolymer (Vicryl, Ethicon Division of Johnson and Johnson Co.); and polydioxanone (PDS, Ethicon Division of Johnson and Johnson Co.).

These aliphatic polyesters are obtained by ring-opening polymerization of the bulk of cyclic diester monomers at high vacuum or under an inert gas atmosphere using stannous catalysts. Sutures are made by the melt extrusion spinning method, followed by a hot-drawing process to give high orientation and crystallinity. After

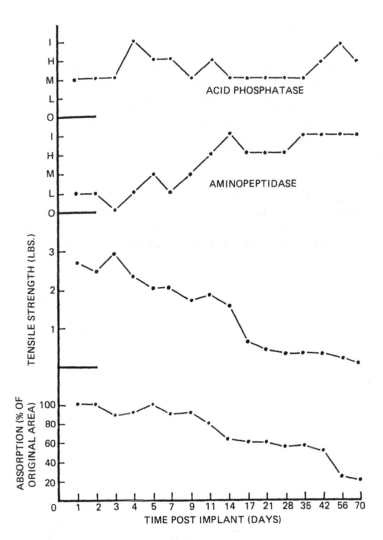

Figure 8.2 Comparison of acid phosphatase and aminopeptidase activity with tensile strength loss and absorption of No. 000 collagen suture. Acid phosphatase activity, denoting the hydrolytic enzyme response, is highest during the first week of implantation, with a later surge of activity when absorption is nearly complete. Aminopeptidase activity is low during the first week, rises to a peak at 2 weeks, and remains high during absorption. Activity was judged visually to arbitrary scale: O, no activity; L, trace; M, moderate; H, increased; and I, intense activity.

this, the sutures are annealed to improve dimensional stability and to prevent extensive shrinkage.

Glycolide Homopolymer (PGA):   Until 1970, only two synthetic suture materials, polypropylene and polyglycolide, had been subjected to a complete premarket clearance by the U.S. Food and Drug Administration (FDA) and received subsequent approval. In 1971, Davis and Geck Division, American Cyanamide Co., marketed the PGA suture for the first time anywhere. The second company that succeeded in commercializing the PGA suture is Japan Medical Supply Co., Hiroshima, Japan, which started sales at the beginning of 1987.

    Figures 3 and 4 show the rate of loss in strength for PGA and catgut sutures after immersion in saline and subcutaneous implantation in rabbits, respectively [7]. One advantage cited for PGA over catgut is the retention of a high percentage of its initial tensile strength during the first 1 to 2 weeks after implantation.

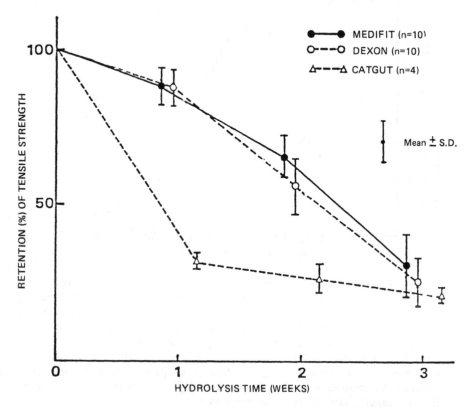

Figure 8.3   Tensile strength change of PGA and catgut sutures as a function of time of hydrolysis in saline (37°C, USP 3-0).

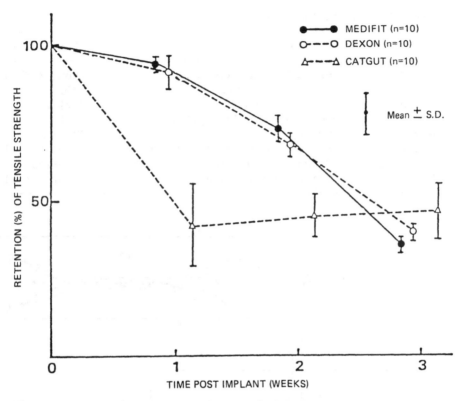

Figure 8.4 Tensile strength change of PGA and catgut sutures as a function of time postimplantation in rabbits (USP 3-0).

Tensile strength and knot security are the two most important mechanical properties of sutures. A knot consists of a number of throws pressed against each other and held in place by frictional contacts. It has been shown that friction is an important parameter and plays a key role in governing the knot security. Gupta et al. determined frictional coefficients of PGA sutures together with non-absorbable suture materials (Table 6) using the testing device shown in Fig. 5 [8]. All of the sutures have the size 00. The tension profiles are compared in Fig. 6 for various sutures. The sutures hold or stick again, then slip, creating a stick-slip effect. The frictional value corresponding to the peak is the static frictional force T_s, and that corresponding to the trough is the kinetic frictional force T_k. The coefficient of friction μ can be calculated using Eq. (1):

Table 8.6 Suture Materials Used

Suture type	Description	Manufacturer
Dexon	Braided polyglycolic acid	Davis & Geck
Ethilon	Monofilament nylon	Ethicon
Mersilene	Braided polyester	Ethicon
Prolene	Monofilament polypropylene	Ethicon
Silk	Braided	Ethicon
Surgilon	Braided silicone-treated nylon	Davis & Geck
Tevdek II	Braided Teflon impregnated polyester	Deknatel
Ticron	Braided silicone-treated polyester	Davis & Geck

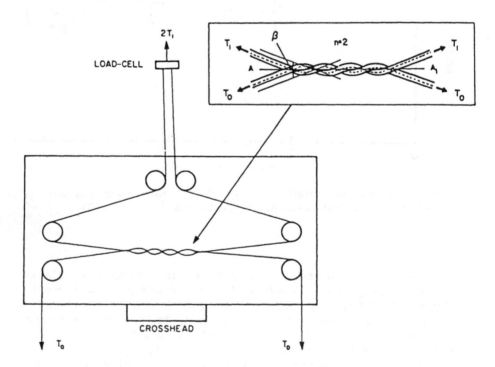

Figure 8.5 The friction testing device. The inset gives the details of the twist geometry.

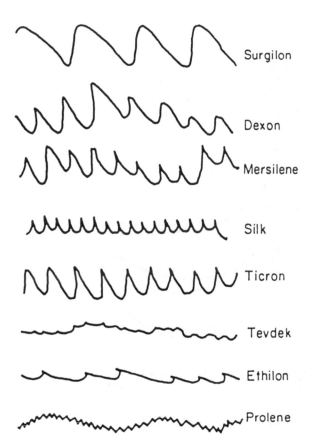

Figure 8.6 Comparison of tension profiles of various sutures (size equals 2-0; n equals 3; T_0 equals 1 pound: full scale load equals 5 pounds, and chart speed equals 5 in./min.

$$\mu = (\pi n \beta)^{-1} (\ln T_1 - \ln T_0) \tag{1}$$

where T_1, T_0, n, and β are the tension activating slippage, the initial tension, the number of turns of the twist, and the twist angle, respectively. The results pertaining to μ_s are plotted in Fig. 7. The value of μ decreases exponentially with the applied tension. It is seen that the coefficient of friction is not a material constant of the sutures but a function of several variables, including applied tension, suture construction, and suture material. Braided sutures such as Dexon give higher frictional values than do the monofilament sutures.

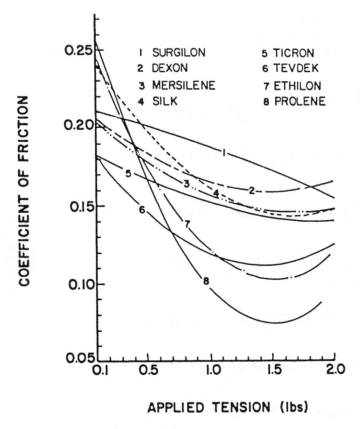

Figure 8.7 Effect of applied tension on the values of the static coefficient of friction of various sutures.

Selection of the best suture material for a given procedure is a subject of lively controversy among surgeons. A material that allows one surgeon to tie secure knots with ease may feel uncomfortable in the hands of another, since he or she may favor the materials with lower frictional coefficients because they pass through tissues with a minimum of drag and are more readily removed. Therefore, a Dexon suture with special surface coating is available (Dexon Plus). Medifit is said to have no coating and to give its good handling by using specially textured yarns.

Healing of tissues closed by suturing is an important question of surgery, especially with bioabsorbable sutures. Recently Lee et al. studied the mechanical strength change with time for anastomosed vessels after surgery [9]. The abdominal aortas and the

carotid arteries of rats were severed, sutured with Dexon, and
then the wounds were closed. After a specific period of time up
to 13 months, the vessels were taken out and tested in uniaxial
loading condition. The maximum forces that a specimen could sustain
are plotted against the age of the suture for the carotid artery and
abdominal aorta in Figs. 8 and 9, respectively. It is seen that in
the first day the maximum force at failure of the sutured artery
is about the same as that of the control. Then the maximum force
decreases with increasing time, until a minimum is reached at 4-6
months. Thereafter the maximum force gains again. At 12 months
the maximum force at failure of the anastomosed arteries rises to
a level that is about the same as that of the control. The stretch
ratio at failure, defined as the overall stretch ratio of the specimen
corresponding to the maximum tensile force sustainable by the specimen, was approximately constant through all periods and had an
average value of 2.1.

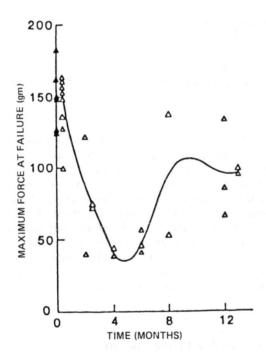

Figure 8.8 Force at failure for carotid artery as a function of
time after operation. Failure loads of the controls at day 0 are
shown in solid black triangles.

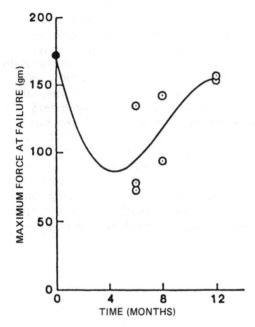

Figure 8.9 Force at failure of abdominal aorta as a function of time after operation. Failure load of the controls at day 0 is shown in solid black circle.

Lee's group also studied the effectiveness of PGA microsutures (Dexon 10-0) on rat uterine anastomoses [10]. Rats were subjected to two-layer (including the mucosa) and one-layer (avoiding the mucosa) interrupted anastomoses of freshly severed uteri as shown in Fig. 10. There was intensive inflammatory reaction during days 7-14 that gradually subsided, and the inflammatory reaction was limited to the suture sites by days 21-30. Sutures were still detectable at the end of 2 months, but they were no longer recognized at the end of 3 months. As the sutures were completely absorbed, giant cells and granulomatous reactions completely cleared. By day 180, the rat uteri in both groups were indistinguishable from non-operated ones.

Such a high effectiveness of PGA suture has also been reported by numerous researchers for other applications. In addition, the Dexon sutures have been compared with nonabsorbable ones. Dorflinger and Kiil compared Dexon and a nonabsorbable suture Dacron, in inguinal and femoral herniorrhaphies [11]. Fifty-eight patients were evaluated at a follow-up examination 6 months after the operation. There was no wound dehiscence or infection during the primary admission. There was one recurrence of hernia in both groups of patients observed for 6 months. As no tensile strength of Dexon was demonstrable after 1 month, hernia recurrence after a longer span of time could not be attributable to the suture type. No fistula or suture granuloma was found in any patient in the 6-month follow-up period, indicating good applicability of absorbable sutures in inguinal and femoral herniorrhaphy. A similar result was reported by Solhaug on repair of inguinal hernia [12]. Comparison was made between Dexon and a nonbiodegradable suture, Mersilene. The respective overall recurrence rates were 5.1% and 4.9%, respectively, as given in Table 7. Persistent neuralgia was more common in the Mersilene group. Suture fistula occurred after one Mersilene repair. This study indicates again that the PGA suture is reliable for preventing recurrent hernia.

The PGA suture was also found to be effective in peripheral nerve anastomosis. Lee et al. studied reanastomoses of freshly severed rat sciatic nerve employing Dexon and nylon 9-0 interrupted epineural suture technique, as shown in Fig. 11 [13]. Animals were

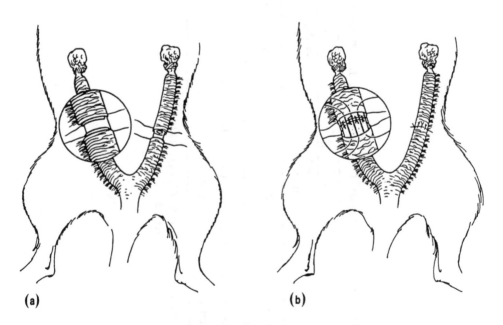

(a)   (b)

Figure 8.10 Rat uterine anastomoses: (a) shows the uterus being reanastomosed with two raw sutures including the mucosa, the left uterus sutured avoiding the mucosa, and (b) shows the mucosa sutured, the seromuscular layers sutured on the right uterus, and the left uterus anastomosed.

Table 8.7 Late Complications (M = Mersilene, D = Dexon)

	1 month n = 520	6 months n = 515	12 months n = 418	3 years n = 258	5 years n = 94
			Time after hernioplasty		
Recurrence[a]	0	12 (7 M, 5 D)	8 (4 M, 4 D)	5 (2 M, 3 D)	1 (1 D)
Neuralgia	20 (11 M, 9 D)	8 (7 M, 1 D)	5 (4 M, 1 D)	2 M[b]	—
Testicular complications	7 (4 M, 3 D)	1 M	—	—	—
Suture fistula	1 M	—	—	—	—

[a]Total recurrence 13/257 M (5.1%), 13/263 D (4.9%).
[b]Nerve resection performed.

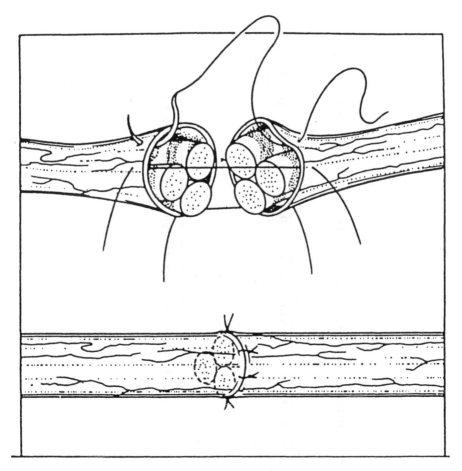

Figure 8.11   Epineural end-to-end sciatic nerve anastomosis.

sacrificed, and sutured portions of the nerves were examined histologically. The results are summarized in Table 8. As is shown, there are almost identical suture reactions by day 7, except that the intensity of polymorphonuclear and macrophage infiltrate in the nylon-sutured nerves was far greater than in the Dexon-sutured group. Although the intensity of the nylon suture reaction subsided considerably after day 21, the lymphocytes, macrophages, and plasma cells surrounded the suture as long as the sutures remained in the nerves, up to 12 months postoperatively. On the contrary, no visible foreign body reaction was detected past the third postoperative month in the Dexon-sutured nerve group. The Dexon sutures were detectable at months 1 and 2 but not after month 3.

Table 8.8 Histological Fate of Peripheral Nerve Anastomoses Using Dexon and Nylon Microsutures (9-0)[a]

Days	7		21		60		90		180	
	Dexon	Nylon	Dexon	Nylon	Dexon	Nylon	Dexon	Nylon	Dexon	Nylon
Inflammatory cell infiltrate	++	+++	++	+++	+	++*	—	+*	—	+*
Necrosis	++	++	+	+	—	—	—	—	—	—
Suture	p	p	p	p	—	p	—	p	—	p

[a]Symbols: +, light; ++ moderate; +++ intense; * inflammatory cells surrounding the suture material only; p suture materials present.

Often, infection takes place when a synthetic material is implanted in the body. Experiments on animals have shown that PGA withstands infection better than most suture materials, except possibly for monofilaments that are noncapillary materials. Capillarity of suture material is claimed to reduce resistance to infections. PGA is even claimed to have an inhibitory effect on microbes due to a hydrolysate, glycolic acid. However, this is unlikely because the acid concentration is too small to play any significant role in practice. Blomstedt made a comparison study on infection between silk and PGA [14]. Silk has been the standard suture material used by many hospitals. He chose silk or PGA at random for 1011 patients operated on in his Department of Neurosurgery, Helsinki University Central Hospital, Finland, during a 19-month period in 1981. One of the results is given in Table 9. The conclusion was that no significant differences in the incidence of serious infections were found between silk and PGA. There was, however, a significant difference in the occurrence of suture fistulas, a matter of inconvenience to the patient, as shown in Table 10. This indicates that PGA or other absorbable suture materials are recommended for closure of the galea, subcutis, fascia, and muscle.

Glycolide/Lactide Copolymer: It was not until the mid-1970s that Ethicon, Inc., brought a competitive biodegradable suture on the market made of a copolymer of glycolide and lactide: the Vicryl (polyglactin 910) suture. The copolymer consists of 90-92% glycolide and 10-8% lactide. The properties seem to be essentially the same as the PGA suture. Table 11 gives the mechanical properties of polyglactin 910, PGA, and other suture materials [15]. However, polyglactin 910 has a relatively high fluid absorption capacity but a low capillary capacity in comparison with PGA. The results are given in Table 12 and Fig. 12, respectively [15]. Evaluations in vivo also give no significant difference between polyglactin 910

Table 8.9   Incidence of Infections Causing Removal of Bone Flap in 556 Patients Randomized for Suture Material

	Suture material	
	Silk	Polyglycolic acid
Bone flap infections	19 (6%)	12 (5%)
Total number of patients with bone flaps	306	250

No significant difference.

Table 8.10 Incidence of Discharging Fistulas and Protruding Stitches in 1011 Patients Randomized for Suture Material

| | Suture material | |
	Silk	Polyglycolic acid
Discharging fistula	17 (3%)	3 (1%)
Protruding stitches	14 (3%)	5 (1%)
Total	31 (6%)	8 (2%)
Total number of patients	543	468

For discharging fistulas p < 0.01 with chi-square with Yates' correction.

Table 8.11 Tensile Strength of Suture Materials Exposed to Blood Plasma for 10 min at 37°C

Materials n = 20 Thread size 2/0	Tensile strength (kp ± S.E.)	Knot tensile strength (kp ± S.E.)	Elongation at break (% ± S.E.)
Polyglactin 910, braided	4.28±0.03	3.44±0.06	19.6±0.1
Polyglycolic acid, braided	4.77±0.17	3.75±0.15	21.4±0.1
Polyamide, monofilament	4.06±0.03	2.86±0.07	43.1±0.7
Polyester braided	3.78±0.02	3.22±0.02	19.6±0.4
Polypropylene, monofilament	2.55±0.04	2.57±0.03	60.9±2.4
Catgut, chromic	1.96±0.09	2.09±0.05	16.7±0.6

Table 8.12 Fluid Absorption of Suture Materials After Exposure to Blood Plasma for 3 Days

Material n = 20 Thread size 0	Fluid absorption in percent of thread dry weight ± S.E.
Polyglactin 910, braided	27.6±0.5
Polyglycolic acid, braided	26.4±0.4
Polyester, braided	16.0±0.3
Polyamide, twisted with cover	27.8±0.8
Linen, twisted	92.2±1.3
Catgut, chromic	101.1±0.8

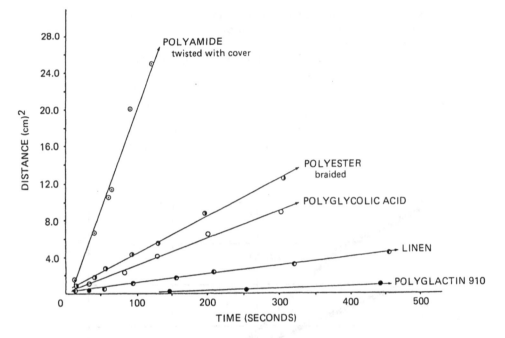

Figure 8.12 Fluid transport rate curves of capillary suture materials. Each point represents the mean of five values of the fluid transport distance (d^2).

and PGA. A result of comparison in layer closure of laparotomy wounds after 306 acute or elective operations is given in Table 13 [16]. The incidence of abscess, granuloma, or sinus formation was 6.5% for PGA and 11.3% for polyglactin 910. The difference is not statistically significant.

Both of these synthetic absorbable sutures show handling properties typical of uncoated braids. They are supple and exhibit high knot security, but tend to pass through tissue less readily than monofilaments and coated braids. Knots must be placed precisely in position, since adjustment of final tension by slippage, as is done with catgut and other smooth-surfaced materials, is difficult. To improve the frictional properties, these synthetic absorbable braided sutures are modified by coating. Coated polyglactin 910 sutures are obtained by coating the braided suture material with a mixture composed of a copolymer of glycolide and lactide and an equal amount of calcium stearate. The breaking strength of the coated Vicryl suture is compared with that of other sutures in Fig. 13 [17]. More than 55% of the original strength remains at 14 days, and over 20% at 21 days.

Table 8.13 Wound Complications in Relation to Acute
and Elective Surgery and Suture Material

	Complication rates in laparotomies using	
Operation	Vicryl	Dexon
Elective	20/107=18.7%	19/102=18.6%
Acute	5/ 44=11.4%	5/ 53= 9.4%

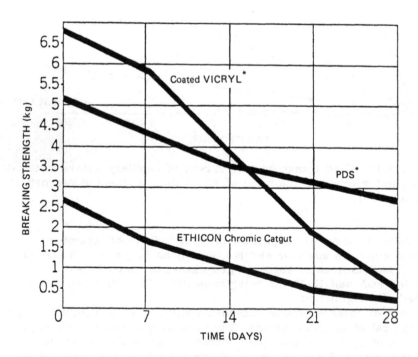

Figure 8.13 Straight pull breaking strength at implantation and
subsequent strength loss of coated Vicryl and other sutures (gauge
3 metric, subcutaneous and intramuscular implants in rats).

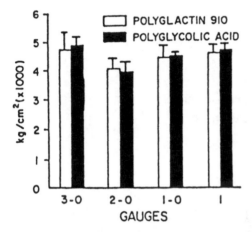

Figure 8.14   Tensile strength of unknotted polyglactin 910 and polyglycolic acid.

The mechanical properties of coated polyglactin 910 and coated PGA (Dexon) were studied by Stone et al. [18]. Again, no significant difference was observed between the two coated materials. Figure 14 shows the tensile strength of unknotted polyglactin 910 and PGA. When the basic 2-1 throw is used, knot security is reached at the same configuration (2-1-1-1) with both materials. Figure 15 demonstrates tensile strengths of implanted polyglactin 910 and PGA with a 2-1-1-1 knot configuration. The polyglactin 910 seems to have greater tensile strength at 21 days. Histologic evaluation revealed no statistically significant difference between the materials.

Polydioxanone: Polydioxanone is obtained by ring-opening polymerization of p-dioxanone:

$$\text{(dioxanone ring, O, C=O)} \rightarrow \left[O\text{-}CH_2CH_2\text{-}O\text{-}CH_2\text{-}\underset{O}{\overset{\|}{C}} \right]_n$$

This is only one monofilament synthetic absorbable suture material, developed by Ethicon, Inc., with a trade name of PDS. PGA and polyglactin 910 can be used as monofilament sutures only in their very finest sizes because of the inherent rigidity of the polymers, which makes them too stiff for general use. A smooth monofilament suture reduces tissue trauma and is easily drawn into place. This is an advantage where a continuous technique is preferred. Another advantage of polydioxanone is to provide significantly longer wound

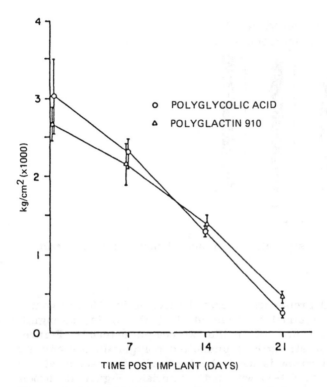

Figure 8.15 Implanted tensile strength of polyglactin 910 and poly-glycolic acid with a 2-1-1-1 knot configuration.

support than other absorbable sutures. The breaking strength change with implantation time is given in Fig. 13 [17]. When implanted in rats, approximately 70% of the suture strength is retained at 2 weeks postimplantation, while at 4 weeks the strength retention is about 50%. At 8 weeks approximately 14% of the original strength remains. The absorption or loss of mass is minimal until about the 90th postimplantation day and is essentially complete within 6 months.

The introduction of a new monofilament and absorbable material has attracted great interest of vascular surgeons and microsurgeons. First experience with the polydioxanone suture in femoro-popliteal bypass was reported by Tuchmann and Dinstl [19]. Angiographical control showed no dilatation or aneurysm formation at the suture line. Histological examination showed fragmented suture material, of which 44% was not yet completely absorbed 168 days postopera-tively. No neutrophilic granulocytes were present, but macrophages, fibrocytes, plasmocytes, and foreign-body giant cells were seen. Tissue reactions were rather slight.

Steen et al. also used the monofilament absorbable suture for arterial anastomoses and compared the result with that of polypropylene suture, one of the most frequently used sutures for vascular surgery today [20]. They made end-to-end anastomoses of the iliac arteries in growing pigs with interrupted 7-0 sutures. The angiographic examination and the recorded tensile breaking forces did not reveal any difference between polydioxanone- and polypropylene-sutured anastomoses. On the other hand, the macroscopic findings, the histological examination, and the blood flow pointed to a far more pronounced inflammatory tissue reaction with polypropylene suture. The calculated blood flow rates are shown in Fig. 16. The rates at the level of anastomoses are 6.1 ± 4.7 and 16.8 ± 2.3 ml/min/100 gr for polydioxanone- and polypropylene-sutures, respectively. These differences clearly suggest the preference of the PDS suture to the Prolene suture for arterial anastomoses. Harjola et al. also found the polydioxanone suture to be applicable to patients treated for lower limb ischemia [21]. The reconstructed and endarterectomized arterial segments were patent at follow-up. The postoperative arteriograms revealed neither aneurysms nor strictures at the anastomoses.

To examine whether or not a suture material associated with a minimal inflammatory response induces less frequent and less severe peritoneal adhesion, Neff et al. compared the adhesion formation between polydioxanone and polyglactin 910 suture in a rabbit model [22]. Sexually mature, virgin female rabbits underwent laparotomy and bilateral incisions into the distal uterine cavities, and were then sacrificed 28 days later. An adhesion score was given as in Table 14. The result is presented in Table 15, where no significant difference is noted in adhesion scores between the two sutures. Similar histologic responses were found in both groups. This study cannot justify the use of one of these sutures over the other with regard to adhesion formation or tissue reaction.

Miscellaneous: As described above, PGA and its derivatives are relatively stiff and inflexible. The rigidity of PGA precludes its use as a monofilament in sizes larger than 7-0. To overcome this problem, Katz et al. incorporated a softer component, trimethylene carbonate (TMC), into the PGA main chain [23]. This copolymer (GTMC) has the chemical structure

$$\{CH_2-\underset{\underset{O}{\|}}{C}-O-CH_2-\underset{\underset{O}{\|}}{C}-O\}_x \quad \{CH_2CH_2CH_2-O-\underset{\underset{O}{\|}}{C}-O\}_y$$

containing approximately 32.5% TMC by weight. Using a standard fiber spinning and subsequent heat setting process, the polymer is extruded into fibers that are processed into monofilament sutures

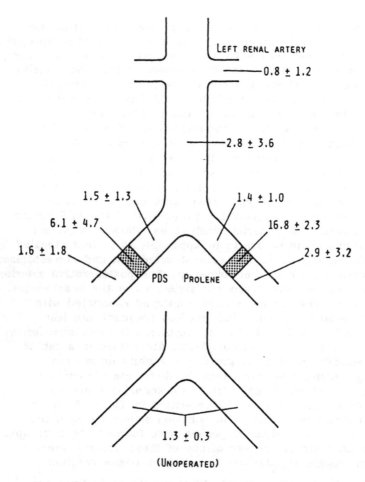

Figure 8.16 Calculated blood flow values (ml/min/100 g) within the walls of the left renal artery, aorta, and iliac arteries at the level of and close to PDS and Prolene anastomoses (n = 6). Also given are the corresponding blood flow values within the walls of the iliac arteries in unoperated control animals (n = 3). Mean ± SD.

Table 8.14 Adhesion Scoring System

Score	Criterion
0	No macroscopic adhesions observed
1	One filmy macroscopic adhesion
2	Multiple macroscopic filmy adhesions
3	One dense macroscopic adhesion with or without additional macroscopic filmy adhesions
4	Multiple dense macroscopic adhesions with or without additional filmy adhesions

Table 8.15 Relationship of Suture Material to Adhesion Score

Suture	Adhesion score				
	0	1	2	3	4
Polydioxanone	1	0	1	6	2
Vicryl	2	1	1	1	5

of various sizes. These sutures have relatively low Young's moduli and high strengths as shown in Table 16 [23]. Decrease in breaking strength in vivo is compared for PGA and this GTMC copolymer in Fig. 17. Percent retention of strength of both materials is similar at 2 weeks, but thereafter, PGA sutures lose strength fairly rapidly: 43% at 3 weeks and about 13% at 4 weeks. Strength retention of GTMC is much greater: 59% at 4 weeks and 30% at 6 weeks. Absorption of the suture was studied in subcutaneous implantation in rabbits. The histologic assessment of absorption is summarized in Table 17 [23]. Apparently, complete absorption occurs between 6 and 7 months in both sizes 00 and 4-0. Untoward tissue reactions were not observed. The results of studies of radiolabeled sutures carried out in the subcutaneous tissues of rats are shown in Fig. 18. It is seen that urine and expired CO_2 are the major excretary routes of the metabolites. Davis and Geck, Inc., has given the trade name Maxon to this new suture.

Longer retention of strength may be needed in some cases, such as in cardiovascular surgery. One candidate material for such a suture is poly-L-lactide (PLA), which is a crystalline polymer similar to PGA but with much lower biodegradation rates. The PLA

Table 8.16 Physical Properties of Glycolide-Trimethylene Carbonate Sutures

Suture size	Material	Knot-pull strength		Straight-pull strength		Elongation to break (%)	Young's modulus (psi)	Diameter (mm)
		(lb_f)	(psi)	(lb_f)	(psi)			
0	GTMC	13.7	54,000	20.6	81,000	33	440,000	0.459
0	Polypropylene	7.5	41,000	11.0	59,000	31	510,000	0.391
0	Nylon	7.5	44,000	11.4	66,000	28	650,000	0.376
00	GTMC	10.8	58,000	14.8	80,000	32	500,000	0.391
00	Polypropylene	5.5	46,000	7.6	63,000	36	420,000	0.314
00	Nylon	6.5	48,000	9.0	67,000	32	590,000	0.333
4-0	GTMC	3.5	57,000	5.4	88,000	38	460,000	0.224
4-0	Polypropylene	2.3	49,000	3.2	67,000	39	660,000	0.198
4-0	Nylon	2.2	54,000	2.9	72,000	31	640,000	0.182
5-0	GTMC	1.4	40,000	3.1	89,000	31	480,000	0.169
5-0	Polypropylene	1.4	54,000	1.7	65,000	36	500,000	0.146
5-0	Nylon	1.3	55,000	1.8	75,000	29	650,000	0.142

GTMC, Glycolide-trimethylene carbonate; psi, pounds per square inch; lb_f, pounds force.

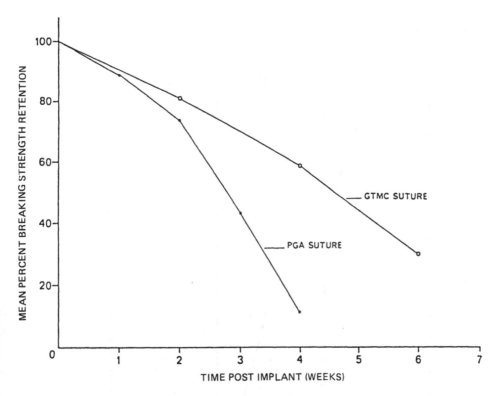

Figure 8.17 Comparison of polyglycolic acid braided sutures and GTMC. The mean retention of strength as percentage of package values after implantation into the subcutaneous tissues of rats is shown.

Table 8.17 Gross Absorption of Glycolide–Trimethylene Carbonate Sutures After Subcutaneous Implantation in the Rabbit

Time (mo.)	No. of rabbits	Suture size 4-0	No. of implants	Suture size 00	No. of implants
3	2	+-++	4	±-+	4
4-1/2	3	+++	6	++	6
6	3	+++-A	6	++-+++	6
7	3	A	6	A	6
8	3	A	6	A	6
9	4	A	8	A	8

±, Active absorption, erosion of suture surface.
+, Less than 50% of suture length absorbed.
++, Fifty percent of suture mass absorbed.
+++, Greater than 50% of suture mass absorbed.
A, Absorbed, a few dye inclusions usually seen.

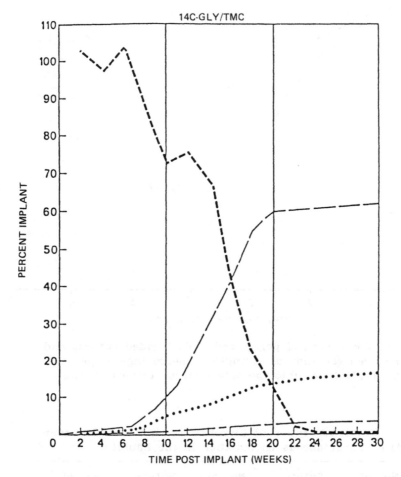

Figure 8.18 Recovery of suture radioactivity and cumulative excretion of radioactivity after implantation of ^{14}C-glycolide-TMC in rats. -----, sutures; — — —, urine; ·····, CO_2; — · —, feces.

fiber can be prepared by a standard melt spinning technique. Figure 19 shows the tensile strengths of the PLA fiber drawn to different elongations and annealed at 140°C [24]. The starting polymer was synthesized under the polymerization conditions described in Table 18. As can be seen in Fig. 19, fibers of higher tensile strengths are obtained as the draw ratio becomes higher. A draw ratio of 6 gives the highest tensile strength, 75 kg·mm^{-2}. X-ray diffraction patterns were diffuse for the undrawn fiber, whereas they changed to very

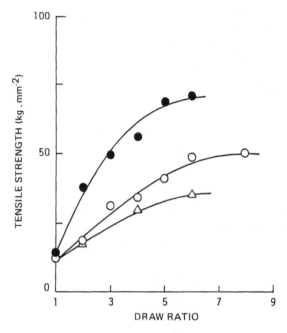

Figure 8.19 Tensile strength of the fiber drawn at different temperatures, followed by annealing at 140°C; (○) 80°C, (●) 160°C, and (△) 80°C for the knotted fiber.

sharp patterns upon drawing and annealing, indicating that highly oriented fibers could be obtained from PLA. The crystallinity estimated from the fiber density was in the range from 5% to 55%. The PLA fiber underwent practically no degradation in phosphate buffer saline at 37°C over 6 months. When the hydrolysis temperature was raised to 100°C, the mass of PLA fiber reduced to half after 30 hr, while Dexon was degraded to the same extent after 2 hr.

Many other synthetic and natural bioabsorbable fibers besides aliphatic polyesters and collagen are currently being studied. Among them are polyamino acids, regenerated collagen, chitin, chitosan, calcium alginate, calcium triphosphate, polyphosphazene, and complex polylactide. Figure 20 shows their light microscopic photographs.

4.2 Fabrics

Little attention has been paid to medical applications of knitted, woven, or nonwoven fabrics obtained from fibers, in comparison with threads.

Table 8.18 Polymerization Conditions and Characteristics
of the Polymer Used for Melt Spinning

Polymerization	
$[\alpha]_{578}^{25}$ (dioxane) of L–lactide	−270°
T_m of L–lactide	97.4°C
[stannous octoate][a]	0.03 wt% of monomer
Polymerization temperature	140°C
Polymerization duration	12 hr
Monomer conversion	97%
Resulting polymer	
$[\alpha]_{578}^{25}$ (dioxane)	−170°
$\overline{M}v$	3.6×10^5
T_g (DSC)	57°C
T_m (DSC)	184°C

[a]Lauryl alcohol was added to the catalyst by 0.01 wt% of
monomer.

An old description is an absorbable gauze from oxidized cellulose
[25]:

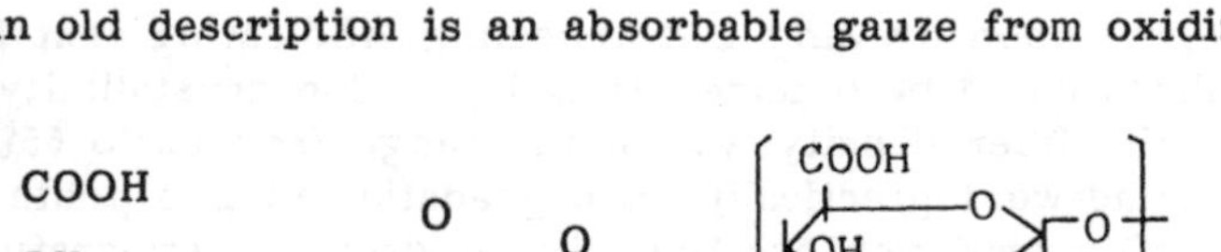

This is currently used mostly as a hemostat in the form of gauze
or cotton. Also, a tricot cloth of oxidized cellulose is applicable
as a hemostat that has a density of 0.03 g·cm^{-2} at lowest, a porosity
around 150 cm^3·sec^{-1}·cm^{-2}, and a liquid absorption capacity of 300%
based on the dry weight [26]. A cloth from lactide polymers is also
claimed to function as a hemostat when the biodegradation rate is
reduced by irradiation with high-energy radiation [27].

A nonwoven fabric of regenerated collagen has been commercialized for wound covering. The collagen fabric, with the trade
name Meipac is made from fibrous atelocollagen obtained by alkaline
treatment of bovine dermis [28].

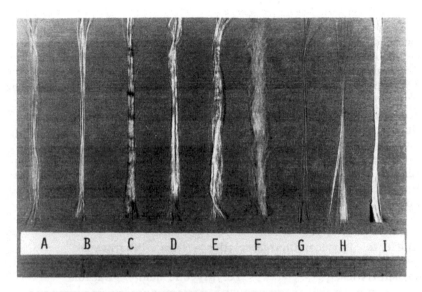

Figure 8.20 Bioabsorbable fibers for medical use under development: A, polyglutamic acid; B, poly(benzyl L-glutamate); C, regenerated collagen; D, chitin; E, chitosan; F, calcium alginate; G, calcium triphosphate; H, polyphosphazene; I, complex polylactide.

Recently, a clinical application of chitin nonwoven fabrics was reported for the treatment of burns [29]. Chitin was purified from shells of Japanese pink crab, dissolved in an amide solvent, and then wet spun into fibers. After cutting to 10 mm, they were dispersed in water in the presence of binder and dried under pressure following water removal. The chitin nonwoven fabric of 10 cm × 10 cm was applied to burns after removal of bulla or debridement of necrotic masses. Fitness to the burn site was very good, and superficial pain was suppressed during epidermalization.

A pledget made of PGA fibers was compared with that of non-biodegradable polytetrafluoroethylene fibers for trachea anastomosis and was concluded to be superior to the nonbiodegradable one .[30]. A tricot cloth of PGA can be used as scaffold for chest wall replacement [31].

4.3 Meshes

At the end of the 1970s, woven meshes made of synthetic absorbable suture materials became available on an experimental basis. Surgical applications of the absorbable meshes have been increasingly reported since then. It is expected that during the absorption period a neomembrane will be formed in the site where the mesh has been present.

Some applications of woven meshes in surgery will be briefly described below. Most of them have been performed using Vicryl (polyglactin 910) meshes. Bowald et al. placed the Vicryl mesh as a patch graft into the thoracic aorta or grafted in the form of a tube to replace a short aortic defect of pigs [32]. They observed that early outgrowth of smooth muscle cells around the mesh took place from the normal aortic media and from a newly formed subintimal smooth muscle layer, the newly formed arterial tissue being completely endothelialized within 20 days, and, in addition, that the polyglactin mesh had disappeared almost completely at 40 days while the new wall seemed to retain sufficient strength throughout the observation time. The arterial regenerative activity was also studied by Greisler et al. by replacing adult rabbit aortas with

Table 8.19 Physical Characteristics of Mesh Prostheses

Characteristics	Polyglycolic acid	Dacron
Threads per inch	56 × 56	60 × 56
Thread-denier (thickness)	230	210
Electrostatic charge	20 V more than Dacron	20 V less than polyglycolic acid
Specific gravity (g/ml)	1.51	1.38
Diameter thread (μm)	150	146.6
Diameter of pore (μm)	304 × 304	304 × 283
Area of pore (mm^2/pore)	0.092	0.08615
Total pore area (mm^2/in.^{2})	288.5	285
Water porosity (ml/cm^2/min at 120 mmHg)	2000	550–650
Water degradable?	Yes	No
Elasticity of thread elongation at break (%)	30.3	22, 26.4, 32
Gain in length after stretch (%)	4	4
Breaking strength (g/d)	6.5	5.7

Table 8.20   Tissue Thickness (mm)

Implantation	TTT	IC	OC
PGA	0.875	0.547	0.378
Dacron	0.262	0.172	0.080

TTT indicates total tissue thickness; IC, inner capsule;
OC, outer capsule; and PGA, polyglycolic acid.

PGA or nonabsorbable Dacron prostheses [33]. Both materials were
woven to similar specifications as shown in Table 19. Between 2
and 4 weeks, circumferentially oriented smooth musclelike myofibro-
blasts proliferated in the PGA inner capsule, yielding a neointima
3.2 times thicker than Dacron's, as shown in Table 20. After 1
month, proliferation stopped in both groups and neointimal thickness
became constant. By 3 months, PGA prosthesis had disappeared
and differentiation of two capsules was no longer possible. All specimens withstood saline infusion at 3 to 5 times systolic pressure.

Defects in a sectional nerve can be bridged using a polyglactin
mesh as described in Fig. 21 [34]. Molandar et al. compared the
nerve repair with conventional nerve grafting in the rabbit and
concluded that the polyglactin tube bridging a fresh nerve defect
of moderate size gave regeneration as good as conventional nerve
grafting [34]. On the other hand, Rosen et al. evaluated a method
of fascicular nerve repair using PGA tube to reconstruct the perineurium and to separate the intrafascicular tissue from the extrafascicular tissue until the nerve established its natural perineurium
[35]. The structure of the peripheral nerve trunk is illustrated
in Fig. 22. The longitudinal orientation of the repairs by fascicular
tubulization was more organized than the repairs by suture, and
the perineurium reestablished more complete continuity in the tube
repairs than in the suture repairs. Quantitative comparisons of
the tube and suture repairs by diameter histograms and axon counts
were not significantly different.

Leakage from the anastomosis after operation often has a high
mortality rate. Arndt et al. applied the Vicryl mesh graft (filament
diameter 0.2 mm and pore size 0.5 mm) for protection of esophagogastrostomy as shown in Fig. 23 [36]. The result is summarized
in Table 21. It appears that protection of an anastomosis with Vicryl
mesh can reduce the tension within and on the anastomotic site and
can achieve adequate sealing of the anastomosis. No anastomotic
dehiscence was seen in the patients, and none died of leakage from
the anastomosis. Splenorrhaphy was also performed in six patients

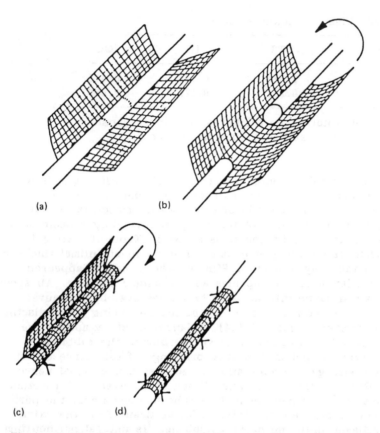

(a)

(b)

(c)

(d)

Figure 8.21   Schematic view of the operative procedure in the polyglactin group: (a) the mesh is placed underneath the nerve and
secured with sutures; (b) a piece of the nerve is resected; (c)
the mesh is shaped like a tube around the nerve; and (d) the tube
is closed by sutures.

using the absorbable Dexon (PGA) mesh [37]. Figure 24 shows
a case of spleen-penetrating injuries from a stab wound. A strip
of PGA mesh was wrapped between the two injuries and sutures
applied through the mesh, incorporating the wounds. The patient
had an afebrile, uneventful postoperative course and was discharged
on the fifth postoperative day. In all six cases, the procedure
was accomplished without abscess formation, postoperative bleeding,
or complications related to the use of the mesh.

In 1909 Albarran described an idea of treating a damaged kidney
by external splinting or encapsulation (Fig. 25) [38]. Schoenenberger

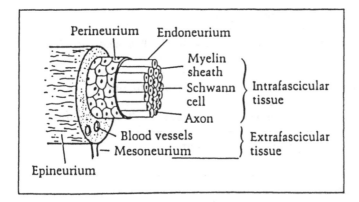

Figure 8.22 Cellular schematic of peripheral nerve trunk. Intrafascicular tissue = endoneurium, Schwann cells, and myelin-axon complexes. Extrafascicular tissue = epineurium and mesoneurium. Perineurium lies at junction of intrafascicular and extrafascicular tissues.

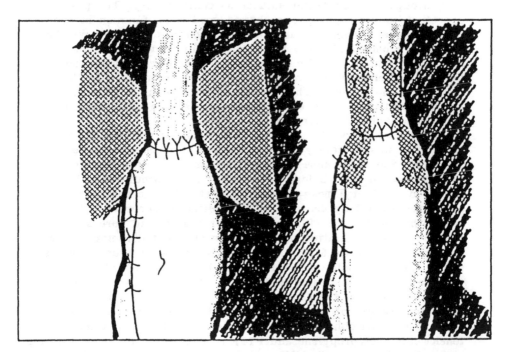

Figure 8.23 Vicryl mesh is passed below the anastomosis, folded around the wound, and fixed with interrupted sutures to the esophagus and stomach, leaving a small anterior gap to prevent stenosis.

Table 21   Results of Esophagogastrostomy

Esophagogastrostomy secured with polyglactin 910 mesh graft. (16 cases)	
7 tumors of the cardia	9 tumors of the esophagus
1 leftthor.	15 abd.-thor.
Dehiscence: 0	periop. mortality: 1, (heart failure)
Esophagogastrostomy without enforcement (34 cases)	
14 tumors of the cardia	20 tumors of the esophagus
2 leftthor.	32 abd.-thor.
Dehiscence: 4	periop. mortality: 3, 2 due to dehiscence

et al. treated experimentally induced blunt renal lesions either with a Vicryl mesh or homologous fibrin adhesive or with through-and-through chromic catgut sutures as controls [39]. The post-operative isotope nephrograms showed good renal function in every case. However, when compared with the controls, the kidneys that had been repaired with the Vicryl mesh contained considerably less scar tissue at the site of parenchymal rupture and showed neither perirenal fibrosis nor atrophy of the parenchyma in the vicinity of the capsule. These results seem to confirm that simple and rapid surgical treatment of moderately severe blunt renal lesions is possible using the Vicryl mesh capsule.

On the contrary, Lamb et al. have reported that the Vicryl mesh is an unsatisfactory material for use for permanent abdominal wall replacement, because adequate fibrous tissue incorporation into polyglactin mesh before hydrolysis did not occur [40]. They compared the Vicryl mesh (filament diameter 140 µm and pore size approximately 400 by 400 µm) with a knitted microporous polytetra-fluoroethylene (PTFE) (filament diameter 140 µm and pore size 500 by 500 µm) and a knitted polypropylene (filament diameter 150 µm and pore size 620 by 620 µm). Full-thickness, inch-square defects in the abdominal walls of rabbits were closed with use of the synthetic meshes as illustrated in Fig. 26. The animals were sacrificed at 3 and 12 weeks, the abdominal walls were removed, and the bursting strength of the grafts and control flaps (vascularized flaps of external oblique fascia) was determined with a tissue tensiometer. Polypropylene and PTFE meshes were similar in bursting strength and not greatly different from controls at 3 and 12 weeks. At 3 weeks polyglactin mesh had a bursting strength comparable to that of control flaps, but at 12 weeks it was significantly weaker.

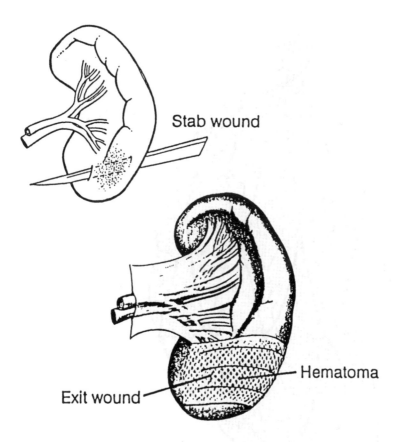

Stab wound

Hematoma

Exit wound

Figure 8.24 Twenty-one-year-old female with stab wound involving the spleen. Repair performed with a portion of mesh incorporating a hematoma and the entry and exit wounds.

Pelvis exenteration leaves a large defect in the floor of the pelvic cavity. A variety of synthetic and biologic materials have been used to create a pelvic lid or floor, but with only limited success. Buchsbaum et al. used a Vicryl mesh to reconstruct the pelvic floor in seven women undergoing pelvic exenteration [41]. The mesh used is a tricot knit derived from Vicryl suture yarn, is 7.5 mil thick, weighs 1.5 oz/yd^2, and has a Mullen burst strength of at least 60 psi. It retained at least 25% of its original strength after 21 days in vivo. The Vicryl mesh was sutured to pelvic peritoneum as shown in Fig. 27. The follow-up ranged from 3 to 31 months, and no patient developed a bowel problem. The material seemed to be appropriate for this use, because it was completely absorbed and acted as a latticework for the deposition of granulation

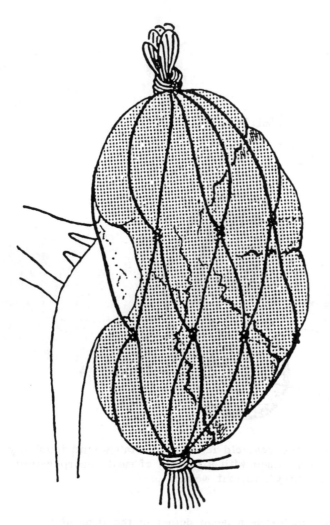

Figure 8.25 Albarran's suggested method (1909) for stabilization of traumatized kidney.

tissue. Maurer and McDonald also have suggested that Vicryl mesh has significant potential as an absorbable, minimally reactive dural substitute [42]. They conducted an experimental project involving closure of dural defects in dogs with a Vicryl mesh graft. Macroscopic and histological examination performed at various times after placement revealed resorption of the graft material, little cerebromembranous adhesion formation, and complete lack of injury to

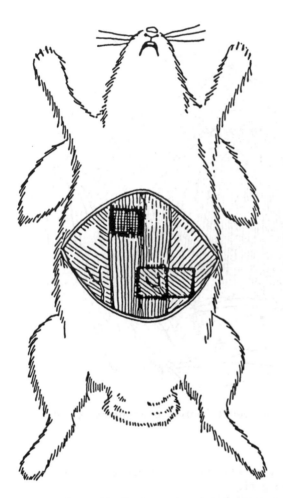

Figure 8.26 Experimental preparation. A 1 in.^{2} area of abdominal wall excised (exclusive of skin) and replaced with a myofascial flap (control) or a piece of mesh.

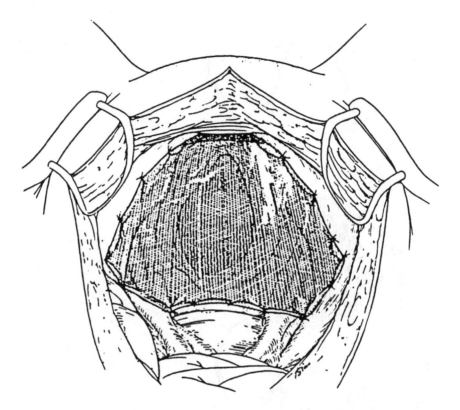

Figure 8.27 Vicryl mesh sutured to pelvic peritoneum, creating diaphragm.

subjacent cortical zones. A substantial neomembrane was formed over the zone of the mesh graft.

Multiple techniques are available for the repair of ruptured human Achilles tendons. Roberts et al. compared (1) unrepaired (conservative cast immobilization), (2) primarily repaired (surgical repair, Fig. 28), and (3) polyglactin mesh-reinforced (Fig. 29) Achilles tendon lacerations in rabbits [43]. All modes of tendon treatment resulted in tensile failure at significantly lower loads than in the contralateral control tendons, but the surgically repaired tendons more closely approximated the load tolerance of the control side than did the unrepaired, casted tendons. The mesh-reinforced tendons exhibited no mechanical superiority to the suture-repaired tendons. No microscopic difference in the quality of collagen deposition for the three types of repairs was apparent.

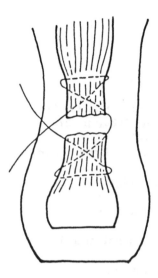

Figure 8.28 Schematic of the Kleinert repair technique for the group managed by suturing only. Tendon fragments were apposed when sutures were tightened.

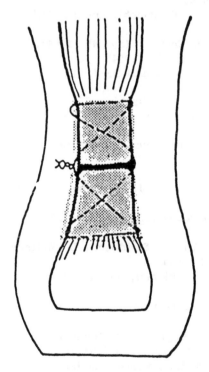

Figure 8.29 Schematic of the repair technique for the group treated by suturing with mesh reinforcement. The approximate extent of mesh coverage is shown by the shaded area.

REFERENCES

1. Capperauld, Suture Biocompatibility, Vth International Conference I. Polymers in Medicine and Surgery, Holland, 1986, p. 27.
2. F. Y. W. Wu, *Obstet. Gynecol.*, *41*, 200 (1973).
3. W. J. Smith, M. A. Johnson, and A. R. Lea, *J. Pharm. Pharmacol.*, *36*, 687 (1984).
4. T. N. Salthouse, J. A. Williams, and D. A. Willigan, *Surg., Gynecol. & Obstet.*, *129*(4), 691 (1969).
5. I. K. Stone, J. A. Fraunhofer, and B. J. Masterson, *Am. J. Obstet. Gynecol.*, 151, 1087 (1985).
6. R. C. Webster, E. G. McCollough, P. R. Giandello, and R. C. Smith, *Arch. Otolaryngol.*, 111, 517 (1985).
7. S. Yamane and T. Tsuchiya, *Jap. J. Artif. Organs*, 15 (4), 1751 (1986).
8. B. S. Gupta, K. W. Wolf, and R. W. Postlethwait, *Surg., Gynecol. Obstet.*, 161, 12 (1985).
9. S. Lee, Y. C. Fung, M. Matsuda, H. Xue, D. Schneider, and K. Han, *J. Biomech.*, *18*(2), 81 (1985).
10. S. Lee, E. Chan, A. R. deMacedo, C. E. Bacchi, and P. Wolf, *Microsurgery*, 5, 15 (1984).
11. T. Dorflinger and J. Kiil, *Acta Chir. Scand.*, *150*, 41 (1984).
12. J. H. Solhaug, *Acta Chir. Scand.*, *150*, 385 (1984).
13. S. Lee, A. R. deMacedo, P. Wolf, K. Han, and S.-J. Li, *Microsurgery*, 4, 120 (1983).
14. G. C. Blomstedt, *Acta Neurochir.*, *76*, 90 (1985).
15. B. Blomstedt and S. I. Jacobsson, *Acta Chir. Scand.*, *143*, 259 (1977).
16. N. Gammelgard and J. Jensen, *Acta Chir. Scand.*, *149*, 505 (1983).
17. Data on file at Ethicon Research Unit.
18. I. K. Stone, J. A. von Fraunhofer, and B. J. Masterson, *Obstet. Gynecol.*, 67 (5), 737 (1986).
19. A. Tuchmann and K. Dinstl, *J. Cardiovasc. Surg.*, 25, 225 (1984).
20. S. Steen, L. Anderson, P. Löwenhielm, H. Stridbeck, B. Walther, and T. Holmin, *Surgery*, *95*(2), 202 (1984).
21. P.-T. Harjola, K. A.-Kulju, and L. Heikkinen, *Thorac. Cardiovasc. Surgeon*, *32*, 100 (1984).
22. M. R. Neff, G. L. Holtz, and W. L. Betsill, *Am. J. Obstet. Gynecol.*, *151*(1), 20 (1985).
23. A. R. Katz, D. P. Mukherjee, A. L. Kaganov, S. Gorden, and P. River, *Surg., Gynecol. Obstet.*, *161*, 213 (1985).
24. S.-H. Hyon, K. Jamshidi, and Y. Ikada, *Polymers as Biomaterials* (S. W. Shalaby, A. S. Hoffman, B. D. Ratner, and

T. A. Horbett, Eds.), Plenum Publishing Corporation, New York, 1984, p. 51.

25. R. Lattes and V. K. Frantz, *Ann. Surg.*, *124*(1), 28 (1946).
26. U.S. Patent 31-657997, October 5, 1984, to Johnson and Johnson Products, Inc.
27. U.S. Patent 31-435148, October 18, 1982, to Johnson and Johnson Products, Inc.
28. U. Takeda, *J. Derm.*, *10*(5), 475 (1983).
29. Y. Ohshima, K. Nishino, Y. Yonekura, M. Maeda, J. Horie, K. Nonomura, S. Kishimoto, T. Wakabayashi, and S. Sotomatsu, *Jap. J. Burn Injuries*, *12*(1), 31 (1986).
30. T. Nakamura, J. Isobe, O. Ike, S. Watanabe, Y. Shimizu, Y. Ikada, S.-H. Hyon, and T. Shimamoto, Experimental Study of Bioabsorbable Pledget for Tracheal Surgery, *J. Jap. Assoc. Thorac. Surg. '86*, Tokyo, 1986, p. 1401.
31. T. Nakamura, S. Watanabe, Y. Shimizu, T. Shimamoto, S.-H. Hyon, and Y. Ikada, Biodegradation of Meshes from Poly-L-lactide and Polyglycolide, Preprint of annual meeting of Japanese Society for Biomaterials '86, Tokyo, 1986, p. 50.
32. S. Bowald, C. Busch, and I. Eriksson, *Surgery*, *86*(5), 722 (1979).
33. H. P. Greisler, D. V. Kim, J. B. Price, and A. B. Voorchees, *Arch. Surg.*, *120*, 315 (1985).
34. H. Molander, O. Engkvist, J. Hägglund, Y. Olsson, and E. Torebjork, *Biomaterials*, *4*, 276 (1983).
35. J. M. Rosen, V. R. Hentz, and E. N. Kaplan, *Ann. Plastic Surg.*, *11*(5), 397 (1983).
36. M. Arndt, B. Lingemann, and B. Kessler, *Acta Chim. Scand.*, *151*, 93 (1985).
37. H. M. Delany, A. Z. Rudavsky, and S. Lan, *J. Trauma*, *25*(9), 909 (1985).
38. J. Albarran, *Médecine operatoire des voies urinaires*, Masson et Cie., Paris, 1909, p. 280.
39. A. Schoenenberger, D. Mettler, H. Roesler, A. Zimmermann, J. Bilweis, W. Schilt, and E. J. Zingg, *J. Urol.*, *134*, 804 (1985).
40. J. P. Lamb, T. Vitale, and D. L. Kaminski, *Surgery*, *93*(5), 643 (1983).
41. H. J. Buchsbaum, W. Christopherson, S. Lifshitz, and S. Bernstein, *Arch. Surg. 120*, 1389 (1985).
42. P. K. Maurer and J. V. McDonald, *J. Neurosurg.*, *63*, 448 (1985).
43. J. M. Roberts, G. L. Goldstrohm, T. D. Brawn, and D. C. Mears, *Clin. Orthopaed. Related Res.*, *181*, 244 (1983).

9

SPINNING THERMOTROPIC POLYMERS

WILLIAM R. KRIGBAUM / Duke University, Durham, North Carolina
Carolina

1. INTRODUCTION

For conventional polymers, such as nylon 6,6 and poly(ethylene
terephthalate), fibers are spun with minimal orientation. According
to Ziabicki [1], orientation of conventional fiber forming polymers
introduced by shear in the spinneret is lost due to relaxation
in the die swell region just below the spinneret. Orientation is
then introduced by cold drawing following spinning. Although the

thermotropic copolyesters such as the Tennessee Eastman X7-G [2-4] and the Hoechst Celanese Vectra copolymers [5,6] have been commercialized as injection molding resins, we [7-12] have investigated the fiber forming properties of these materials. A number of other workers [13-16] have studied the fiber properties of the Tennessee Eastman X7-G copolymers.

Only a moderate chain extension is required to form a nematic phase in the melt. However, a long relaxation time is essential so that orientation introduced in the spinning process is not lost before the fiber crystallizes. As Jackson [17] has shown, highly extended chains are necessary for a long relaxation time. A highly extended chain implies a low entropy of fusion and a high melting temperature. Thus, some randomness must be introduced to lower the melting temperature so that processing can occur without degradation. Workers at du Pont have patented several compositions involving random substitution of groups such as phenyl on the rings of hydroquinone [18] or terephthalic acid [19] in substituted poly(p-phenylene terephthalate). However, these polymers have not been commercialized. In two of the polymers that have been commercialized, the Tennessee Eastman X-7G and the Hoechst Celanese Vectra copolyesters, randomness is introduced by copolymerization. Flory [20] has pointed out that melt interchange of a copolymer in a one phase system leads to a random copolymer. However, as Lenz et al. [21] have demonstrated, melt interchange in a two phase system can lead to a blocky polymer. The second phase, which allows blocks to develop, can be either a crystalline or a nematic phase. Blockiness significantly affects the rheological properties of the melt and hinders orientation by spin drawing. This produces fibers that have poor mechanical properties.

Our study [7] began with a sample of Tennessee Eastman's X7-G, a copolymer of poly(ethylene terephthalate) having 60 mol% poly(p-hydroxybenzoate). This polymer will be designated as PET/60PHB. Thermotropic polymers spun from the nematic phase cannot be cold drawn, so orientation must be introduced by shear in the capillary or by spin drawing the molten fiber just below the spinneret. Spin draw is defined as the ratio of the take-up velocity of the fiber to the extrusion velocity of the melt at the spinneret, V_f/V_o. The spun fiber was collected with a take-up machine located 1.0 m below the capillary. since the die swell ratio for nematic polymers is near unity [14,22,23], it may be possible to introduce orientation by shear in the spinneret. During our study [7] we learned that this sample was blocky. Figure 1 shows the differential scanning calorimeter (DSC) heating curve of the blocky PET/60PHB sample as received. There are melting transitions at 200 and 260°C, and other transitions at 227 and 320°C. Figure 2 shows a log-log plot

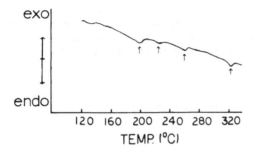

Figure 9.1 DSC of a blocky sample of PET/60PHB recorded at 20°C/min. (From ref. 7.)

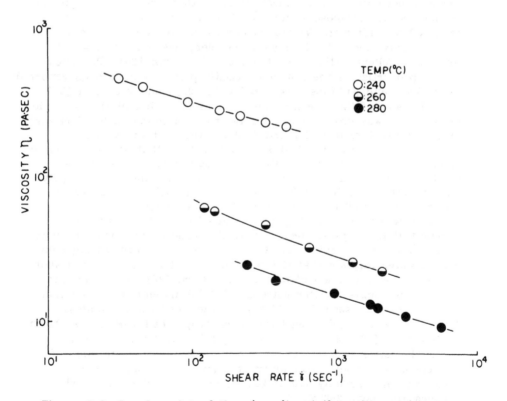

Figure 9.2 Log-log plot of the viscosity at three temperatures of a blocky sample of PET/60PHB shown as a function of shear rate at the wall. (From ref. 7.)

of the melt viscosity versus shear rate at temperatures 240, 260, and 280°C. The viscosity is shear thinning over this 40° temperature range. We note that the largest drop in melt viscosity occurs between 240 and 260°C. This is reasonable in view of the 260°C melting temperature of this copolymer.

2. RHEOLOGICAL STUDIES OF THE TENNESSEE EASTMAN COPOLYMER

Rheological investigations were started early in the study of thermotropic polymers. At that time the only sample available in significant quantities was the Tennessee Eastman X-7G. For this reason the PET/60PHB and PET/80PHB copolymers were studied. We have recently performed a rheological study of three different compositions of the Hoechst Celanese Vectra copolymers. We will discuss these results later, after the Vectra copolyesters have been introduced.

The rheology of liquid crystal polymers has been reviewed by White and Fellers [24], Baird [25], and Wissbrun [26]. The rheological properties of thermotropic nematic polymers have been reported by Jackson and Kuhfuss [2-4], Baird [27,28], and Wissbrun [22]. The viscosity of the nematic phase is less than that of the isotropic phase. This was demonstrated for lyotropic mesophases by Hermans in 1962 [29]. Jerman and Baird [30] showed that the viscosity of the nematic phase of PET/60PHB was less than that of nonmesogenic PET. Wissbrun [31] was the first to demonstrate the difference for the nematic and isotropic phases of the same thermotropic polymer. Gotsis and Baird [32] measured the viscosity of the nematic phase of PET/60PHB over a 10^6 sec^{-1} range of shear rate. The viscosity was shear thinning over most of the range, but became constant at the highest shear rates. The rheo-optical data of Onogi et al. [33] for lyotropic cholesteric solutions of hydroxypropylcellulose in water indicate that the relaxation of stress occurs in a shorter time than the relaxation of orientation. Baird [28] made transient flow measurements for thermotropic PET/60PHB and concluded that stress relaxes in seconds, while orientation relaxes in minutes. Also, die swell is quite small for thermotropic PET/60PHB, as shown by Wissbrun [22], Jerman and Baird [23], and Sugiyama et al. [14]. The latter two references demonstrate that the die swell of this copolymer is less than unity at temperatures below 260°C, becoming approximately 1.0 at 260°C, and increasing slowly with temperature above 260°C.

3. THE EFFECT OF SHEAR IN THE CAPILLARY

Our first question is whether shear in the capillary, or elongational flow introduced by spin drawing, is the more effective way to orient

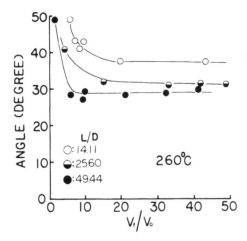

Figure 9.3   X-ray crystallite orientation angle shown for three
L/D ratios versus spin draw ratio V_f/V_0. (From ref. 7.)

thermotropic copolymers. We have spun [7] PET/60PHB at temperatures between 240 and 280°C using the Sieglaff-McKelvey rheometer.
The capillaries furnished with this instrument all have the same
length. However, we can increase shear in the capillary by reducing
the diameter of the capillary, thereby increasing the length-to-
diameter ratio, L/D. Figure 3 illustrates the X-ray orientation angle
for fibers spun at 260°C versus the spin draw ratio for three different L/D ratios. A lower orientation angle corresponds to better
orientation of crystallites in the fiber. As L/D is increased, the
crystallite orientation improves at a lower spin draw ratio. Figure 4
shows the initial modulus as a function of the spin draw ratio for
fibers spun through capillaries differing in the length-to-diameter
ratio. As the L/D ratio is increased, which corresponds to more
shear in the capillary, the modulus increases and less spin draw
is required to obtain the plateau value. However, shear in the
capillary has no effect if there is no spin draw. This is probably
due to the parabolic flow profile in the capillary and the flat profile
in the crystallized fiber, as suggested by Ide and Ophir [34].
The outer part of the fiber solidifies more rapidly, so that spin
draw accelerates the sheath of the fiber and helps to convert the
parabolic profile into a flat profile. According to Sugiyama et al.
[14], no spin draw is required in the spinning of X-7G. They find
that shear in the capillary is sufficient to give maximal crystalline
orientation in the spun fiber. We observe that, contrary to their
conclusion, some spin draw is essential to obtain good properties.
We suspect that their fibers were allowed to free fall, so that gravity
provided some spin draw.

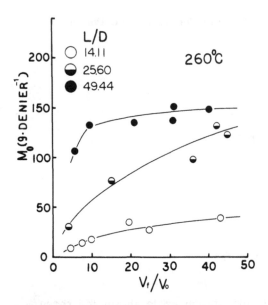

Figure 9.4 Initial modulus of a blocky sample of PET/60PHB for three L/D ratios shown as a function of the spin draw ratio. (From ref. 7.)

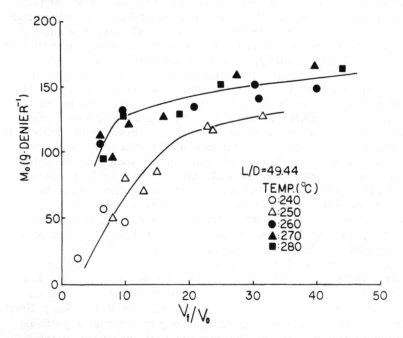

Figure 9.5 Initial modulus of fibers spun at five temperatures from a blocky sample of PET/60 PHB as a function of spin draw ratio. (From ref. 7.)

Figure 5 shows the initial modulus versus spin draw ratio for five spinning temperatures. These fibers are spun with a capillary having L/D = 49.44, so the spin draw ratio is rather small. Poor properties are obtained at spinning temperatures of 240 and 250°C, but a single curve fits the data for spinning temperatures of 260, 270, and 280°C. This observation is expected in view of the DSC melting temperature of 260°C. The maximum initial modulus is only 150 g/denier, which is rather low.

4. ORIENTATION INTRODUCED BY ELONGATIONAL FLOW DUE TO SPIN DRAWING

We next reduced the L/D ratio from 49.44 to 14.1 and spun fibers at higher spin raw ratios [9]. Figure 6 shows the initial modulus of fibers spun at higher spin draw ratios. The spinning temperatures are 250, 260, 280, and 300°C. The best fibers have an initial modulus over 200 g/denier, as compared with 150 g/denier for fibers spun at high L/D. This indicates that spin draw is more effective in orienting the fiber than shear in the capillary. In Fig. 6 we note a difference between spinning temperatures 260, 280, and 300°C. This result differs from that obtained at high L/D, because in that case the initial modulus of fibers spun at 260, 270 and 280°C fell on the same curve. In view of the DSC melting temperature of 260°C, it is surprising that the initial modulus improves between 280 and 300°C. We believe that at high spin draw ratios the fiber becomes thinner and cools more rapidly, so that crystallization is nucleated in the threadline by higher melting PHB crystallites created by blockiness. To test this hypothesis, we need a more sensitive method than DSC to detect the presence of a small amount of crystallinity caused by short poly(p-hydroxybenzoate) (PHB) blocks.

Cogswell [35], and Wissbrun and Ide [36], have reported an interesting effect of thermal history on the rheology and spinning properties of these thermotropic polyesters. The melt viscosity is reduced, and the fiber properties are improved, if the melt is heated to a higher temperature, T_H, and then cooled to the spinning temperature, T_S. We next tested whether preheating would have an effect on the rheology at temperatures above the DSC melting temperature of 260°C. Figure 7 shows the melt viscosity versus shear rate for samples tested at 280, 300, and 280°C after preheating the melt to 300°C (filled circles). We note that the filled circles nearly fall on the 300°C curve, indicating that some crystallinity is destroyed between 280 and 300°C. This provides the sensitive test we sought for the presence of small amounts of crystallinity.

Figure 8 shows the X-ray orientation angle versus spin draw ratio for fibers spun at 250°C, 280°C, and for fibers spun at 250°C

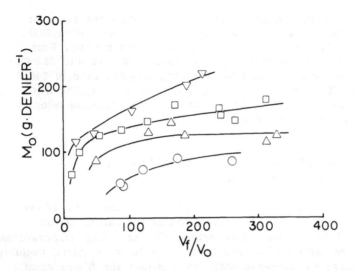

Figure 9.6 Initial modulus of fibers spun from blocky PET/60PHB at low L/D versus spin draw ratio. Spinning temperatures are 300°C (▽), 280°C, (□), 260°C (△), and 250°C (○). (From ref. 9.)

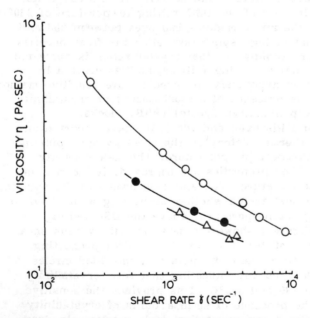

Figure 9.7 Log-log plot of melt viscosity versus apparent shear rate for blocky PET/60PHB at temperatures of 280°C (○), 300°C (△), and at 280°C, after preheating the melt to 300°C (●).

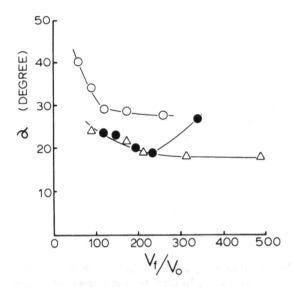

Figure 9.8 Crystallite orientation angle alpha plotted as a function of spin draw ratio for fibers spun at 250°C (o), at 280°C (△), and for fibers spun at 250°C after preheating the melt to 280°C (●). (From ref. 9.)

after preheating the melt to 280°C. Note that the preheated sample follows the 280°C curve to a spin draw ratio of 234, but then increases to the 250°C curve. This indicates that, with increasing spin draw ratio, the crystallite orientation becomes poorer.

Figure 9 illustrates the initial modulus of fibers spun at 250°C, 280°C, and for fibers spun at 250°C after preheating the melt to 280°C. The filled circles representing the preheated sample fall on the 280°C curve until the spin draw ratio exceeds 234, and then they decrease to the 250°C curve. At higher spin draw ratios, the fiber becomes thinner and cools faster. This decrease is probably due to crystallization in the threadline, which will be demonstrated below.

DSC curves are shown in Fig. 10 for fibers spun at 250°C (top three curves). These show crystalline melting endotherms at 260°C and 270°C at all three spin draw ratios. The middle set of three curves represent fibers spun at 280°C. These show no melting transitions in the 260°C range. The lower set of four curves represent fibers spun at 250°C after preheating the melt to 280°C. Melting endotherms appear at 265°C for spin draw ratios 234 and above, which is the region in which the X-ray orientation angle shown in Fig. 8 increases and the initial modulus shown in Fig. 9 falls to the 250°C curve.

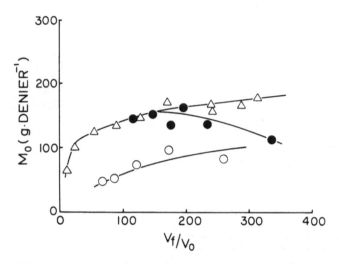

Figure 9.9 Initial modulus of fibers spun from blocky PET/60PHB as a function of spin draw ratio for spinning temperatures of 250°C (o), 280°C (△), and at 250°C after preheating the melt to 280°C (●). (From ref. 9.)

The initial modulus is illustrated in Fig. 11 for fibers spun at 280, 300, and at 280°C after preheating the melt to 300°C (filled circles). The latter points fall very near the 300°C spinning temperature curve, indicating that preheating gives the fiber modulus expected at the preheating temperature. However, as the spin draw is increased beyond 140, the initial modulus falls to the value expected for the spinning temperature of 280°C. Higher spin draw ratios thin the fiber, allowing it to cool more quickly, so that crystallization occurs in the threadline. We conclude that a very small amount of HBA crystallinity reduces the ability to orient the polymer by spin draw, because these crystallites nucleate crystallization in the threadline.

Figure 12 indicates that the decrease in initial modulus can be postponed by increasing the temperature below the spinneret. These temperatures are 49, 42, and 35°C. At 49°C the higher modulus extends to somewhat higher spin draw ratios, while at 35°C the initial modulus always falls on the lower curve. This indicates that crystallization in the threadline prevents good orientation of the polymer. Acierno and co-workers [13] found that the initial modulus of PET/60PHB spun at high spin draw ratios increased with decreasing spinning temperature. This result was ascribed to reinforcement by crystallites. We conclude, on the contrary,

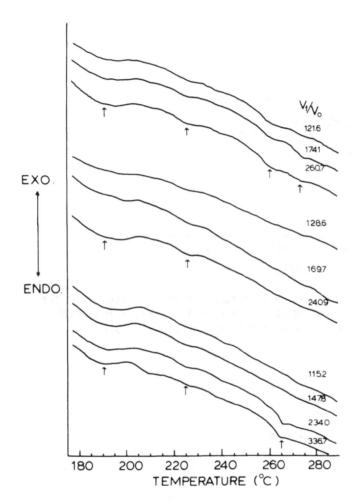

Figure 9.10 DSC curves taken at 10°C/min for fibers at different spin draw ratios spun at 250°C (upper three curves), at 280°C (middle three curves), and at 250°C after preheating the melt to 280°C (lower four curves). (From ref. 9.)

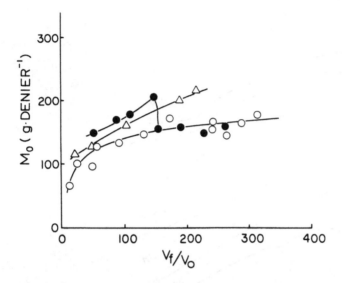

Figure 9.11 Initial modulus as a function of spin draw ratio for spinning temperatures of 300°C (△), 280°C (○), and at 280°C after preheating the melt to 300°C (●). (From ref. 9.)

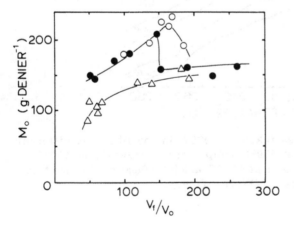

Figure 9.12 Initial modulus versus spin draw ratio for fibers spun at 280°C after preheating to 300°C. Temperatures 7 cm below the spinneret are 49°C (○), 42°C (●), and 35°C (△). (From ref. 9.)

that at high spin draw ratios the presence of crystallites prevents the attainment of good crystallite orientation and a high value of the initial modulus.

5. COMPARISON OF THE BEHAVIOR OF BLOCKY AND RANDOM COPOLYMERS

The rheological and spinning behaviors described above were clarified [10] when we received a second sample from Tennessee Eastman that was nearly random. These two copolymers had the same inherent viscosity, but one was blocky as received (OLD) and the other was nearly random (NEW). We might expect samples of the PET/60PHB copolymer to be random from the work of Jackson and Kuhfuss [2] and McFarlane et al. [37]. However, samples of this copolymer are produced in small quantities, so that we may anticipate some variation from sample to sample. Figure 13 compares the DSC heating

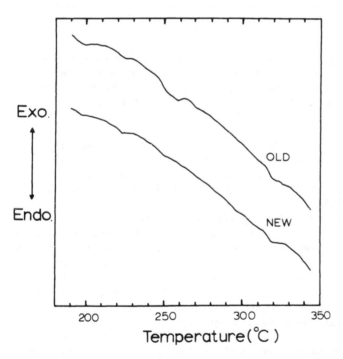

Figure 9.13 DSC heating curves taken at 10°C/min for the blocky OLD sample (above) and the more random NEW sample (below) of PET/60PHB. (From ref. 10.)

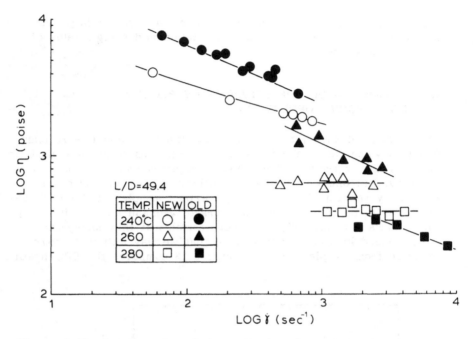

Figure 9.14 Log-log plot of the melt viscosity versus apparent shear rate for three temperatures. Values for the more random NEW sample are designated by open symbols and the blocky OLD sample of PET/60PHB by filled symbols. (From ref. 10.)

curves of these two samples. The upper curve, representing the blocky copolymer, exhibits a melting endotherm at 200°C and a large endotherm at 260°C. The more random sample shown in the lower curve has a melting endotherm at 200°C and a very small endotherm at 250°C.

The difference in a log-log plot of the viscosity of the nematic phase versus shear rate is shown in Fig. 14 for these two samples. The filled symbols represent the more blocky sample. Its viscosity decreases with increasing shear rate at all three temperatures. The more random sample, represented by open symbols [10], shows Newtonian behavior at temperatures of 260°C and above. The melt rheology for these two samples is quite different. In the latter temperature range the random sample has no crystallinity, and its viscosity is independent of shear rate. This suggests that the shear thinning behavior observed for the more blocky sample can be attributed to the presence of HBA crystallites at the three temperatures.

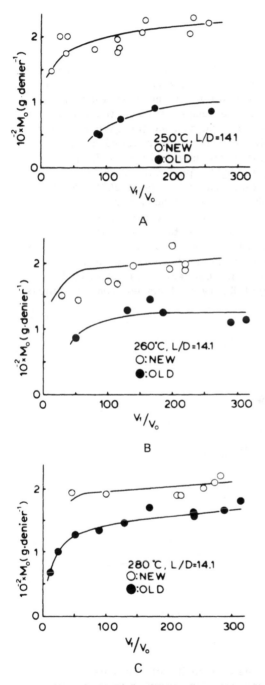

Figure 9.15 Initial modulus versus spin draw ratio for spinning temperatures of (a) 250°C, (b) 260°C, and (c) 280°C. Open symbols indicate the NEW sample and filled symbols the OLD sample of PET/60PHB. (From ref. 10.)

Figure 15 compares the initial modulus of fibers spun at low
L/D ratio and higher spin draw for the blocky (OLD) and the more
random (NEW) samples. The initial modulus of the more random
sample [10] is independent of spinning temperature in the range
250-280°C. Also, fibers from the more random sample give a higher
initial modulus, 220 g/denier. This difference is reduced as the
spinning temperature of the blocky copolymer is increased, but
some difference remains for a spinning temperature of 280°C. This
suggests that some HBA crystallinity remains at least 20°C above
the DSC melting endotherm at 260°C for the blocky sample.

A log-log plot of melt viscosity versus shear rate is shown
in Fig. 16 for the more random sample. The test temperatures are
260, 280, and 260°C after preheating the melt to 280°C (filled circles)
[10]. Preheating has no effect on the melt viscosity of the more
random sample. This demonstrates that preheating is effective in
removing the small amount of crystallinity that arises from blockiness.

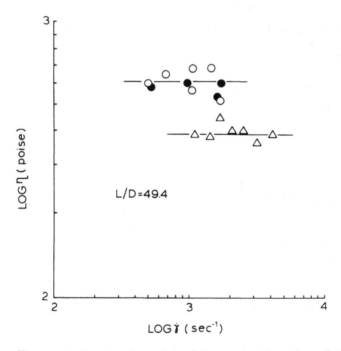

Figure 9.16 Log-log plot of the melt viscosity of the more random
NEW sample versus apparent shear rate for spinning temperatures
of 260°C (o), 260°C after preheating the melt to 280°C (●), and
at 280°C (ᴧ). (From ref. 10.)

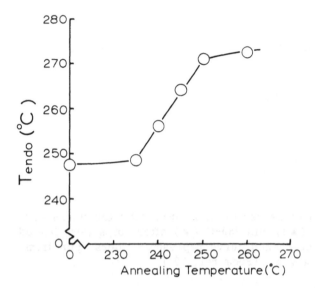

Figure 9.17 Temperature of the melting endotherm after chips of the random NEW sample were heated for 1 hr at the indicated temperature. (From ref. 10.)

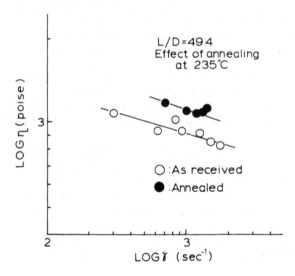

Figure 9.18 Melt viscosity measured at 250°C versus apparent shear rate for the random NEW sample (open circles) and after chips of the NEW sample were heated for 1 hr at 235°C (filled circles). (From ref. 10.)

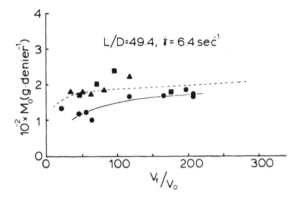

Figure 9.19 Initial modulus versus spin draw ratio for fibers spun at 280°C (■), 260°C (▲), and 250°C (●) after chips were heated for 1 hr at 235°C. The dashed curve represents fibers spun from the NEW sample as received. (From ref. 10.)

We next heated different samples of chips of the more random copolymer for 1 hr at a series of increasing temperatures [10]. Figure 17 illustrates the increase in the DSC melting temperature when chips of the random copolymer are heated for 1 hr at the indicated temperatures. These results indicate that melt interchange transforms the random copolymer into a blocky copolymer. Note that the melting temperature is hardly increased when the chips are heated for 1 hr at 235°C. However, we will see that this is sufficient to change the rheology and spinning behavior.

Figure 18 compares [10] the melt viscosity measured at 250°C of the more random sample as received (open circles) and of the sample heated for 1 hr at 235°C (filled circles). Heating for 1 hr at 235°C, which leads to a small amount of blockiness, has increased the melt viscosity.

The initial modulus of fibers versus spin draw ratio at spinning temperatures of 250, 260, and 280°C after the chips were heated for 1 hr at 235°C is shown [10] in Fig. 19. The initial modulus of fibers spun at 250°C falls below the dashed line, representing the initial modulus of fibers spun from untreated chips. The small amount of crystallinity introduced by heating for 1 hr at 235°C is melted out at 260°C, so the initial modulus of fibers spun at these temperatures is unchanged by this minimal heat treatment.

6. RHEOLOGY OF THE HOECHST CELANESE VECTRA COPOLYESTERS

The Hoechst Celanese Vectra copolyesters have various compositions of p-hydroxybenzoic acid (HBA) and 2-hydroxy-6-naphthoic acid

(HNA). We were provided three samples having the compositions 75HBA/25HNA. 58HBA/42HNA, and 30HBA/70HNA [38]. The highest inherent viscosities of these three samples were 9.2, 5.1, and 7.8 dL/g, and their melting temperatures, as measured by DSC, are 295, 247, and 309°C, respectively. All three samples exhibited a yield stress, indicating the presence of crystallites at the test temperatures [28]. Melt fracture is related to flow instability. When melt fracture occurs, the diameter of the spun fiber varies along the length of the fiber. Melt fracture can be initiated in two ways: (a) inlet melt fracture at the entry to the capillary [39-42], and (b) slip-stick flow at the capillary wall [43-45]. In the case of inlet melt fracture, there is a critical shear stress at the onset of flow instability. If the critical stress is exceeded, the surface becomes rougher. When the flow instability occurs at the capillary wall, melt fracture occurs at low shear stress and disappears when the critical stress is exceeded.

7. PROPERTIES OF FIBERS SPUN FROM VECTRA COPOLYESTERS

We have investigated the properties of fibers spun from Vectra co-polyesters having the compositions 75HBA/25HNA [11], 58HBA/42HNA [8], and 30HBA/70HNA [12]. The 58HBA/42HNA copolymer is near the eutectic composition, and provides better fiber properties than the other two samples.

We obtained three samples of the 75HBA/25HNA copolymer having inherent viscosities 3.0, 6.0, and 9.2 dL/g. The melting temperatures of these three samples as measured by DSC were 281, 284, and 295°C. This suggests that the samples of higher inherent viscosity were produced by melt interchange of the sample having the lowest inherent viscosity. We anticipate that this process will produce longer HBA blocks. These polymer samples exhibited melt fracture, so it was necessary to exceed the critical stress to obtain fibers with a uniform diameter. The critical stress increased with increasing inherent viscosity and with decreasing spinning temperature. We selected 320°C as the spinning temperature. The sample having inherent viscosity 9.2 dL/g reached a maximum initial modulus of 112 g/denier at a spin draw ratio of only 40. The polymer having inherent viscosity 3.0 dL/g gave a maximum initial modulus of 208 g/denier at a spin draw ratio of 108. In the case of the copolyester produced by Tennessee Eastman, PET/60PHB, the crystallite orientation paralleled the behavior of the initial modulus. However, the crystallite orientation of 75HBA/25HNA improved monotonically with spin draw ratio, while the initial modulus reached a maximum and then decreased. We conclude that increasing the spin draw ratio for 75HBA/25HNA introduces defects in the partially crystalline fiber, which reduce the mechanical properties.

The 58HBA/42HNA copolyester exhibited DSC endotherms at 222 and 247°C, an indication of two crystal polymorphs. This polymer was spun at 250, 260, and 280°C. The initial modulus was low for fibers spun at 250°C. The initial modulus versus spin draw fell on a single curve for spinning temperatures of 260°C, 280°C, and for fibers spun at 250°C after preheating the melt to 280°C. The initial modulus of 58HBA/42HNA fiber spun with an L/D ratio of 49.4, and a spin draw ratio of 130, was approximately 420 g/denier. This can be compared with 200 g/denier when PET/60HBA was oriented by spin draw. This difference can be ascribed to the fact that PET units make the chain more flexible than HNA units.

We next spun a sample of 30HBA/70HNA having inherent viscosity 7.8 dL/g. This sample exhibits shear thinning viscosity and yield stress at low shear stress, which indicates that crystallites are present at the test temperatures. The yield stress decreases with increasing temperature. The viscosity is significantly reduced if the blocky copolymer is preheated to a higher temperature and then cooled to the test temperature. The melting temperatures of this sample, as measured by DSC, were 304 and 323°C, and the spinning temperatures were 325 and 335°C. In order to obtain smooth fibers and avoid melt fracture, a shear rate of 1030 sec^{-1} was used at 325°C and 630 sec^{-1} at 335°C. The mechanical properties of fibers spun at the lower temperature were clearly inferior to those spun at 335°C, and the crystallite orientation was poorer. Fibers spun at 325°C exhibited a maximum initial modulus of approximately 70 g/denier at a spin draw ratio of only 14. The initial modulus of fibers spun at 335°C reached an optimum value of 185 g/denier at a spin draw ratio of 43. This behavior parallels that of the 75HBA/25HNA copolymer. As the composition departs more from the eutectic ratio, we can anticipate a more blocky nature in the synthesized copolymer. Melt interchange will then lead to longer blocks in copolymers far from the eutectic composition.

8. IMPROVING FIBER PROPERTIES BY HEAT TREATMENT OF FIBERS

Luise [46] has disclosed a procedure to improve the mechanical properties by heat treating spun fibers. Heat treatment raises the crystalline melting temperature, so each stage involves a higher temperature, but always below the current crystal melting temperature. Our interest is not to determine the optimum heat treatment conditions, but to find what properties are responsible for the improvement in mechanical behavior. For this reason we have heat treated spun fibers of the Hoechst Celanese Vectra copolymers

58HBA/42HNA [8] and 30HBA/70HNA [45], using a single temperature for each composition. DSC heating curves of the 58HBA/42HNA
copolymer indicate two crystalline polymorphs with melting temperatures of 222 and 247°C. Fibers of the 58HBA/42HNA polymer were
heat treated at 231°C. Fibers spun with smaller spin draw ratios
have larger diameters, so the increase in molecular weight, the
crystalline melting temperature, and the degree of crystallinity
with heat treatment time occurs more slowly. However, if the
58HBA/42HNA fiber spun with spin draw ratio 5.8 is heat treated
for a longer time, it will show the same increase in molecular weight,
crystal-nematic melting temperature, and degree of crystallinity
as a fiber with spin draw ratio 68. This is illustrated in Fig. 20,
where the increase of inherent viscosity is shown as a function
of heat treatment time for spin draw ratios of 5.8 and 68. Figure 21
illustrates the increase of initial modulus for fibers spun from the
58HBA/42HNA copolymer as a function of heat treatment time for
fibers spun with different spin draw ratios. There is no improvement, even for long heat treatment times, if the spin draw ratio
is 36 or lower. Fiber mechanical properties improve more rapidly
with heat treatment time as the spin draw ratio increases. DSC
endotherms for the 30HBA/70HNA copolymer provided evidence of
two crystal polymorphs at 304° and 323°C. Fibers of this composition
were heat treated at 305°C. Similar results were obtained for both
compositions, but the mechanical properties of the 30HBA/70HNA
copolyester showed improvement at a spin draw ratio of 32.

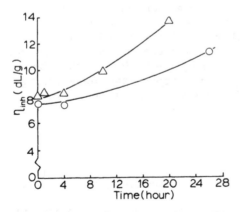

Figure 9.20 Inherent viscosity versus heat treatment time for fibers
with spin draw ratios 68 (△) and 5.8 (○). (From ref. 8, copyright © 1986; reprinted by permission of the American Chemical
Society.)

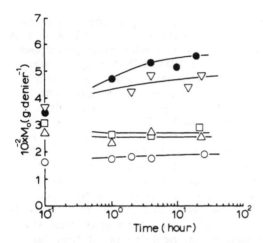

Figure 9.21 Initial modulus versus heat treatment time for spin
draw ratios 68 (●), 44 (▽), 36 (□), 12.2 (▵), and 5.8 (○).
(From ref. 8, copyright © 1986; reprinted by permission of the
American Chemical Society.)

Heat treatment of fibers improves their mechanical properties
only if the fibers are well oriented due to a high spin draw ratio.
Fibers prepared with a low spin draw raio, if heated for a longer
time, will exhibit the same increase in molecular weight, crystal
melting temperature, and degree of crystallinity. However, these
increases are not paralleled by an increase of tenacity or initial
modulus. This suggests to us that these three properties are not
related to the improvement of fiber tenacity or initial modulus.
We suspect that the interface between the two crystal forms provides
many defects that reduce the fiber properties. Heating the fibers
at successively higher temperatures will replace the lower melting
polymorph with the higher melting crystal form, thereby eliminating
the interface between the two crystal forms.

REFERENCES

1. A. Ziabicki, in *Fundamentals of Fiber Formation*, Wiley, London,
   1976, p. 200.
2. H. F. Kuhfuss and W. J. Jackson, Jr., U.S. Patent 3,778,410,
   1973.
3. H. F. Kuhfuss and W. J. Jackson, Jr., U.S. Patent 3,804,805,
   1974.

4. W. J. Jackson, Jr., and H. F. Kuhfuss, *J. Polym. Sci. Polym. Chem. Ed.*, *14*, 2043 (1976).
5. G. W. Calundann, U.S. Patent 4,161,470, 1979.
6. G. W. Calundann, U.S. Patent 4,184,996, 1980.
7. H. Muramatsu and W. R. Krigbaum, *J. Polym. Sci. Polym. Phys. Ed.*, *24*, 1695 (1986).
8. H. Muramatsu and W. R. Krigbaum, *Macromolecules*, *19*, 2850 (1986).
9. H. Muramatsu and W. R. Krigbaum, *J. Polym. Sci. Polym. Phys. Ed.*, *25*, 803 (1987).
10. H. Muramatsu and W. R. Krigbaum, *J. Polym. Sci. Polym. Phys. Ed.*, *25*, 2303 (1987).
11. W. R. Krigbaum, C. K. Liu, and D. K. Yang, *J. Polym. Sci. Polym. Phys. Ed.* (in press).
12. D. K. Yang and W. R. Krigbaum, *J. Polym. Sci. Polym. Phys. Ed.* (in press).
13. D. Acierno, F. P. La Mantia, G. Polizzotti, A. Ciferri, and B. Valenti, *Macromolecules*, *15*, 1455 (1982).
14. H. Sugiyama, D. N. Lewis, J. L. White, and J. F. Fellers, *J. Appl. Polym. Sci.*, *30*, 2329 (1985).
15. A. Tealdi, A. Ciferri, and G. Conio, *Polym. Comm.*, *28*, 22 (1987).
16. J. A. Cuculo and G. Y. Chen, *J. Polym. Sci. Polym. Phys. Ed.*, *26*, 179 (1988).
17. W. J. Jackson, Jr., *Br. Polym. J.*, *12*, 154 (1980).
18. C. R. Payet, U.S. Patent 4,159,365, 1979.
19. J. F. Harris, Jr., U.S. Patent 4,294,955, 1981.
20. P. J. Flory, in *Principles of Polymer Chemistry*, Cornell University Press, Ithaca, N.Y., 1953, chap. III.
21. R. W. Lenz, J. I. Jin, and K. A. Feichtinger, *Polymer*, *24*, 327 (1983).
22. K. F. Wissbrun, *Br. Polym. J.*, *12*, 163 (1980).
23. R. E. Jerman and D. G. Baird, *J. Rheol.*, *25*, 275 (1981).
24. J. L. White and J. F. Fellers, *J. Appl. Polym. Sci. Appl. Polym. Symp.*, *33*, 137 (1978).
25. D. G. Baird, in *Liquid Crystal Order in Polymers* (A. Blumstein, Ed.), Academic, New York, 1978, pp. 237-259.
26. K. F. Wissbrun, *J. Rheol.*, *25*, 619 (1981).
27. D. G. Baird, in *Liquid Crystal Order in Polymers* (A. Blumstein, Ed.), Academic, New York, 1978, pp. 237-259.
28. D. G. Baird, in *Polymeric Liquid Crystals* (A. Blumstein, Ed.), Plenum, New York, 1980, pp. 119-143.
29. J. Hermans, *J. Colloid Sci.*, *17*, 638 (1962).
30. R. E. Jerman and D. G. Baird, *J. Rheol.*, *25*, 275 (1981).

31. K. F. Wissbrun, paper presented at AIChE Meeting, Houston, 1983.
32. A. D. Gotsis and D. G. Baird, *J. Rheol.*, *29*, 539 (1985).
33. Y. Onogi, J. L. White, and J. F. Fellers, *J. Non-Newt. Fluid Mech.*, *7*, 121 (1980).
34. Y. Ide and Z. Ophir, *Polym. Eng. Sci.*, *23*, 261 (1983).
35. F. N. Cogswell, *Br. Polym. J.*, *12*, 170 (1980).
36. K. F. Wissbrun and Y. Ide, U.S. Patent 4,325,903, 1982.
37. F. E. McFarlane, V. A. Nicely, and T. G. Davis, *Contemp. Top. Polym. Sci.*, *2*, 109 (1977).
38. D. K. Yang and W. R. Krigbaum, *J. Polym. Sci. Polym. Phys. Ed.* (in press).
39. J. P. Tordella in *Rheology, Theory and Applications* (F. R. Eirich, Ed.), Vol. 5, Academic Press, New York, 1969, chap. 2.
40. J. P. Tordella, *Trans. Soc. Rheol.*, *1*, 203 (1957).
41. E. B. Bagley and A. M. Birks, *J. Appl. Phys.*, *31*, 556 (1960).
42. A. E. Everage, Jr., and R. L. Ballman, *J. Appl. Polym. Sci.*, *18*, 933 (1974).
43. J. J. Benbow, R. V. Charley, and P. Lamb, *Nature*, *192*, 223 (1961).
44. A. V. Ramurthy, *J. Rheol.*, *30*, 337 (1986).
45. Y. H. Lin, *J. Rheol.*, *29*, 605 (1985).
46. R. R. Luise, U.S. Patent 4,247,514, 1981.

INDEX